This important and unique collection captures a living experiment in the Nile Basin. Ethiopia shattered the basin's status quo by launching construction of the Grand Ethiopian Renaissance Dam, Africa's largest hydroelectric project. Important precedents in managing transboundary resources may be set as Egypt, Sudan and Ethiopia thrash out sharing benefits as well as water. The contributors address the legal issues, the regional politics and the projected economic impact of Ethiopia's unilateral action.

John Waterbury, *President Emeritus,*
the American University of Beirut, Lebanon

This important volume examines the impacts of the GERD project through an interdisciplinary lens, incorporating insights from the fields of law, political science, economics and hydrology. As the contributors show, the game-changing nature of the GERD may introduce a new era of cooperation on the Nile.

Stephen McCaffrey, *Distinguished Professor of Law, University of the Pacific,*
McGeorge School of Law in Sacramento, USA and 2017 Laureate,
Stockholm Water Prize.

Cooperative management of transboundary fresh waters is a great challenge of our time – nowhere more so than in the Nile basin, with eleven riparian nations and the world's longest river. This scholarly book makes an invaluable contribution at a turning point in Nile history. The authors, many with long experience studying Nile issues, have woven together an important interdisciplinary study of the risks and opportunities arising from the construction of the Grand Ethiopian Renaissance Dam on the Blue Nile, linking Ethiopia, Sudan and Egypt.

David Grey, *Visting Professor of Water Policy,*
University of Oxford, UK

The Grand Ethiopian Renaissance Dam and the Nile Basin

The Grand Ethiopian Renaissance Dam (GERD) will not only be Africa's largest dam, but it is also essential for future cooperation and development in the Nile River Basin and East African region. This book, after setting out basin-level legal and policy successes and failures of managing and sharing Nile waters, articulates the opportunities and challenges surrounding the GERD through multiple disciplinary lenses.

It sets out its possibilities as a basis for a new era of cooperation, its regional and global implications, the benefits of cooperation and coordination in dam filling, and the need for participatory and transparent decision making. By applying law, political science and hydrology to sharing water resources in general and to large-scale dam building, filling and operating in particular, it offers concrete qualitative and quantitative options that are essential to promote cooperation and coordination in utilising and preserving Nile waters. The book incorporates the economic dimension and draws on recent developments including: the signing of a legally binding contract by Egypt, Ethiopia and Sudan to carry out an impact assessment study; the possibility that the GERD might be partially operational very soon; the completion of transmission lines from GERD to Addis Ababa; and the announcement of Sudan to commence construction of transmission lines from GERD to its main cities. The implications of these are assessed and lessons learned for transboundary water cooperation and conflict management.

Zeray Yihdego is a Reader in International Law at the University of Aberdeen, UK.

Alistair Rieu-Clarke is a Professor of Law at the University of Northumbria, UK.

Ana Elisa Cascão is currently an Independent Researcher/Consultant, and until recently was a Programme Manager at Stockholm International Water Institute, Sweden.

Earthscan Studies in Water Resource Management

Community Management of Rural Water Supply
Case Studies of Success from India
Paul Hutchings, Richard Franceys, Stef Smits and Snehalatha Mekala

Drip Irrigation for Agriculture
Untold Stories of Efficiency, Innovation and Development
Edited by Jean-Philippe Venot, Marcel Kuper and Margreet Zwarteveen

Water Policy, Imagination and Innovation
Interdisciplinary Approaches
Edited by Robyn Bartel, Louise Noble, Jacqueline Williams and Stephen Harris

Rivers and Society
Landscapes, Governance and Livelihoods
Edited by Malcolm Cooper, Abhik Chakraborty and Shamik Chakraborty

Transboundary Water Governance and International Actors in South Asia
The Ganges-Brahmaputra-Meghna Basin
Paula Hanasz

The Grand Ethiopian Renaissance Dam and the Nile Basin
Implications for Transboundary Water Cooperation
Edited by Zeray Yihdego, Alistair Rieu-Clarke and Ana Elisa Cascão

Freshwater Ecosystems in Protected Areas
Conservation and Management
Edited by Max C. Finlayson, Jamie Pittock and Angela Arthington

For more information and to view forthcoming titles in this series, please visit the Routledge website: www.routledge.com/books/series/ECWRM/

The Grand Ethiopian Renaissance Dam and the Nile Basin

Implications for Transboundary Water Cooperation

**Edited by Zeray Yihdego,
Alistair Rieu-Clarke and
Ana Elisa Cascão**

LONDON AND NEW YORK

First published 2018
by Routledge

2 Park Square, Milton Park, Abingdon, Oxfordshire OX14 4RN
52 Vanderbilt Avenue, New York, NY 10017

Routledge is an imprint of the Taylor & Francis Group, an informa business

First issued in paperback 2019

British Library Cataloguing in Publication Data
A catalogue record for this book is available from the British Library

Library of Congress Cataloging in Publication Data
A catalog record for this book has been requested

ISBN: 978-1-138-06489-8 (hbk)
ISBN: 978-0-367-37690-1 (pbk)

Typeset in Goudy
by Wearset Ltd, Boldon, Tyne and Wear

Contents

Notes on contributors   ix
Preface   xiii
List of abbreviations   xvi

1 A multi-disciplinary analysis of the risks and opportunities of
   the Grand Ethiopian Renaissance Dam for wider cooperation
   in the Nile   1
   ZERAY YIHDEGO, ALISTAIR RIEU-CLARKE AND
   ANA ELISA CASCÃO

2 The Nile Basin Cooperative Framework Agreement:
   disentangling the Gordian Knot   18
   SALMAN M.A. SALMAN

3 Agreement on declaration of principles on the GERD:
   levelling the Nile Basin playing field   41
   SALMAN M.A. SALMAN

4 International law developments on the sharing of Blue Nile
   waters: a fairness perspective   61
   ZERAY YIHDEGO AND ALISTAIR RIEU-CLARKE

5 Changing cooperation dynamics in the Nile Basin and the
   role of the GERD   90
   ANA ELISA CASCÃO AND ALAN NICOL

6 GERD and hydropolitics in the Eastern Nile: from
   water-sharing to benefit-sharing?   113
   RAWIA TAWFIK AND INES DOMBROWSKY

7 **Analysing the economy-wide impacts on Egypt of alternative GERD filling policies** 138

BRENT BOEHLERT, KENNETH M. STRZEPEK AND
SHERMAN ROBINSON

8 **Economic impact assessment of the Grand Ethiopian Renaissance Dam under different climate and hydrological conditions** 158

TEWODROS NEGASH KAHSAY, ONNO KUIK,
ROY BROUWER AND PIETER VAN DER ZAAG

9 **From projecting hydroclimate variability to filling the GERD: upstream hydropower generation and downstream releases** 181

YING ZHANG, SOLOMON TASSEW ERKYIHUN
AND PAUL BLOCK

10 **Managing risks while filling the Grand Ethiopian Renaissance Dam** 193

KEVIN G. WHEELER

Index 216

Contributors

Paul Block is a Professor at the University of Wisconsin-Madison in Civil and Environmental Engineering, focusing on hydroclimatology and water resources systems management. He has been engaged in Ethiopia-related projects since 2003, dating back to his graduate studies.

Brent Boehlert is a Principal at Industrial Economics, Inc. (IEc) and a Research Affiliate with the Massachusetts Institute of Technology (MIT). He holds a PhD in Water Resources Engineering and an MSc in Natural Resource Economics. Dr Boehlert has worked in over twenty countries, recently serving as an advisor to the government of Zimbabwe on their National Water Master Plan and analysing the contribution of water resources development to Uganda's economy. His experience in the Nile Basin focuses on the economic implications of climate change and cooperative water management.

Roy Brouwer is Professor in the Department of Economics at the University of Waterloo, Canada, where he is also Executive Director of the Water Institute. He is a visiting professor at the Swiss Federal Institute for Aquatic Science and Technology (Eawag) in Zurich. He obtained his PhD in Environmental Economics from the University of East Anglia, UK.

Ana Elisa Cascão has been an Independent Researcher and Consultant since September 2017. Previously, she was a Programme Manager at the Stockholm International Water Institute (SIWI). The research published in this book has been done through SIWI's research centre, the International Centre for Water Cooperation (a UNESCO CAT II Centre). Ana Elisa Cascão joined SIWI in February 2010 as a Programme Manager and has worked across several different units – Applied Research, Capacity Building, Advisory Services and lately at the Transboundary Water Management Unit – and in several different regions such as the Nile Basin, East and Southern Africa and the Middle East/North Africa. Her main research field is hydropolitics and transboundary water resources management/cooperation in the Eastern Nile Basin, and she is the author of several academic publications on these topics.

Ines Dombrowsky holds a PhD in Economics and an MSc in Environmental Engineering and heads the Department 'Environmental Policy and Natural Resources Management' at the German Development Institute/Deutsches Institut für Entwicklungspolitik (DIE). Her research is grounded in institutional and ecological economics, political sciences and human geography and focuses on natural resources and multilevel environmental governance, with a particular focus on water and the water–energy–food–climate nexus. She has published widely on water governance issues from transboundary to local scales. Prior to joining the DIE in September 2010, she was a researcher at the Helmholtz Centre for Environmental Research – UFZ in Leipzig, Germany (2001–2010). She has also had practical work experience working on transboundary water cooperation in the Nile Basin with the World Bank from 1997 to 2001 and on the Jordan River with the Deutsche Gesellschaft für technische Zusammenarbeit (GTZ) from 1995 to 1997 respectively.

Onno Kuik is a Researcher at the Institute for Environmental Studies of the VU University Amsterdam in the Netherlands. He received his PhD from the economic faculty of VU University on a study about climate policy and international trade. He is an experienced CGE modeller.

Tewodros Negash Kahsay is an Assistant Professor in the Department of Economics at Addis Ababa University. He received his MSc from Agricultural University of Norway (now Norwegian University of Life Sciences) in development and resource economics and his PhD from the Department of Environmental Economics at VU University Amsterdam in economic analysis of transboundary water management.

Alan Nicol is Principal Researcher and Sustainable Growth Program lead at the International Water Management Institute, based in Addis Ababa, Ethiopia. His work focuses on the intersection of institutions, knowledge, development and power with respect to water resources management at different scales. He has lived and worked in Nile Basin countries since the 1990s.

Alistair Rieu-Clarke is an International Environmental Law Professor at Northumbria Law School, Newcastle, UK. He holds a Bachelor of Laws (Honours), a Masters in Natural Resources Law and Policy (with distinction), and a PhD in international law, sustainable development and transboundary waters. Alistair Rieu-Clarke has almost twenty years' experience in research, teaching, professional training and consultancy activities across Africa, Asia, Europe and South America. These activities primarily focus on exploring law's contribution to transboundary water conflict and cooperation.

Sherman Robinson is a Senior Research Fellow at the International Food Policy Research Institute (IFPRI) and Professor of Economics (emeritus) at the University of Sussex. Professor Robinson is a leading expert on global and national economic simulation models, particularly computable general equilibrium (CGE) models, which have become a standard tool of analysis of the economic impact of climate change, trade and fiscal policy, regional integration, structural adjustment, and development strategies. Before joining IFPRI in 1993, he was Professor of Agricultural and Resource Economics at the University of California, Berkeley (1983–1994); Economist, Senior Economist and Division Chief in the Research Department of the World Bank (1977–1983); Assistant Professor of Economics at Princeton University (1971–1977); and Lecturer in Economics at the London School of Economics (1969–1971). He has been a consultant to the World Bank and has held visiting senior-staff appointments at the Economic Research Service, U.S. Department of Agriculture; the U.S. Congressional Budget Office; and the President's Council of Economic Advisers (in the Clinton administration).

Salman M.A. Salman is the Editor-in-Chief of the *International Water Law* journal, and a Fellow with the International Water Resources Association (IWRA). Until 2009, he served as the World Bank Adviser on Water Law and Policy. Prior to that, he worked as a Legal Officer with the International Fund for Agricultural Development of the United Nations (IFAD) in Rome, Italy, and taught law at the University of Khartoum in Sudan. Dr Salman has published extensively on issues related to water law and policy, which can be accessed at www.salmanmasalman.org

Kenneth M. Strzepek is a Research Scientist at MIT's Joint Program on the Science and Policy of Global Change, Adjunct Professor of Public Policy at Harvard Kennedy School of Government, and Non-Resident Senior Fellow at the UNU World Institute for Development Economics Research. He has a PhD in Water Resources Systems Analysis from MIT, and an MA in Economics from the University of Colorado. Professor Strzepek was an Arthur Maass-Gilbert White Fellow at the Institute for Water Resources of the US Army Corps of Engineers and received the US Department of Interior Citizen's Award for Innovation in the applications of Systems Analysis to Water Management. He has been working on the water and agricultural issues of the Nile Basin since 1977, most recently focusing on using hydro-economic and economy-wide modelling to assist in policy analysis.

Solomon Tassew Erkyihun is a Postdoctoral Research Associate working with water resources systems and society research group under the Civil and Environmental Engineering department of the University of Wisconsin – Madison.

Rawia Tawfik is an Assistant Professor at the Faculty of Economics and Political Science at Cairo University. She holds a Doctor of Philosophy in Politics from the University of Oxford and a Master of Science in Politics from Cairo University. Her research interests include issues surrounding African political economy. She was a visiting research fellow at the South African Institute of International Affairs (2003), the Africa Institute of

South Africa (2009–2010) and the German Development Institute (2015). Her postdoctoral research focuses on hydropolitics in the Nile Basin. She has a number of published research papers, book chapters, and journal articles on Nile hydropolitics, regional integration in Africa, and Egypt's foreign policy towards Africa.

Kevin G. Wheeler is a Doctoral Researcher at the Environmental Change Institute at the University of Oxford and the principal of Water Balance Consulting LLC. He has provided extensive negotiation support for the Colorado River Basin, as well as model development and training for stakeholders throughout the Eastern Nile Basin. He holds a DPhil and MSc in Water Science, Policy and Management from the University of Oxford and an MSc and BSc in Water Resource Engineering from the University of Colorado.

Zeray Yihdego is a Reader in International Law at the University of Aberdeen, School of Law. He has widely researched and written in various aspects of international law including: arms trade law, humanitarian law and international water law. He is involved in a major EU funded Water–Food–Energy Nexus research project relating to the Zambezi and Omo River Basins. He holds PhD and Master of Laws from the University of Durham and the University of Cambridge respectively, both in pubic international law. He also held a Visiting Research Fellow position with the Institute for Ethics, Law and Armed Conflict (ILAC), University of Oxford and a Senior Visiting Member at Linacre College, University of Oxford. He is a member of the UN Expert Group on the regulation of firearms and also acts as in-house consultant for the UN. Dr Yihdego serves as an Editor-in-Chief of the *Ethiopian Yearbook of International Law* (Springer).

Pieter van der Zaag is Professor of Water Resources Management at IHE Delft and at Delft University of Technology, The Netherlands, and has led several multidisciplinary research projects on agricultural water management, water allocation and on transboundary river basin management, mainly in Africa.

Ying Zhang is a Postdoctoral Fellow at Johns Hopkins University currently (2017) after receiving her PhD in Civil and Environmental Engineering from University of Wisconsin – Madison. Her work is focused on water resource system engineering and management.

Preface

The final statement of the Budapest Water Summit 2016 recognised water as: a crucial factor for attaining the Sustainable Development Goals (SGDs) 2015, a critical natural asset that requires protection, and an enabler and an inter-connector of the various targets of the SDGs related to poverty, food, energy, health and climate change. A key recommendation from the Summit was to encourage countries to 'establish or revitalize appropriate local, regional, national, basin-level and transboundary institutions that address the allocation and sustainable use of water in a fair, transparent and equitable manner' (Budapest Water Summit 2016). This reinforces one of the targets of the SDGs, which is to, 'implement integrated water resources management at all levels, including through transboundary cooperation as appropriate' (SDGs 2015: Goal 6). An associated indicator for this SDG, upon which the progress of states will be measured, is the proportion of a transboundary basin area with an 'operational arrangement' for water cooperation.

The need for effective cooperation with respect to the use and management of the world's more than 263 international river basins is widely recognised in the international community. These river basins provide 60 per cent of the word's freshwater flow, are home to 40 per cent of the world's population, and offer huge potential for human development. However, if their use and protection is not managed through cooperation, they can also be a source of unhealthy competition and tension among states and communities. Such competition, if unchecked, can be exacerbated by the drastic increase in world population, the uncertainties brought about by climate change, and the influence of other geo-political and socio-economic factors. Therefore, a responsible, sustainable and equitable approach to the governance of these river basins has become critical now more than ever.

The Nile Basin is a typical example of the challenges and opportunities that are evident when sharing and managing transboundary water resources. More than 487 million people, or 40 per cent of the population of the African continent, live in the Nile Basin. Most of the eleven upstream Nile riparian countries are classified as 'least developed countries' (LDCs) that are eager to get their populations out of extreme poverty. The two furthest downstream riparian countries – one of which is classified as poor (Sudan) while the other classified as a middle-income country (Egypt) – rely primarily on the Nile and related water resources for their national political economies and the livelihoods of their populations. While the two downstream Nile riparians make the most use of Nile waters, upstream riparians have increasingly recognised and capitalised on the potential to make further use the Nile waters. Managing such potentially conflicting positions through cooperation is considered to be beneficial to both upstream and downstream interests.

In the last decade, the Nile Basin countries have made significant progress in reconciling their differences and advancing towards basin-wide governance. Under the umbrella of the Nile Basin Initiative (NBI), they have managed to agree on a common vision of achieving 'sustainable socio-economic development through the equitable utilization of, and benefit from, the common Nile Basin water resources'. The NBI, as a transitional cooperative arrangement, has not yet resulted in an agreement by *all* riparians on key legal principles, procedures and a permanent institutional framework for the Nile Basin. The Cooperative Framework Agreement (CFA) signed and ratified by six and four countries respectively is still a source of contention. In turn, the lack of a permanent legal and institutional framework is one of the impediments to cooperation, and perhaps to achieving one of the targets of the SDGs 2015.

In this context, the Grand Ethiopian Renaissance Dam (GERD), which is under construction on the Blue Nile (one of the main sub-systems of the Nile) in Ethiopia close to the Sudanese border, should be seized as an opportunity and potential catalyst for greater cooperation among (Eastern) Nile Basin countries.

Following on from, and building upon, the Special Issue published with *Water International* (WI) in August 2016, this volume contains an important collection of papers devoted to the GERD. It is a multidisciplinary product of an international consortium of academics and practitioners growing out of a special session at the International Water Resources Association (IWRA) XV World Water Congress in Edinburgh 25–29 May 2015. Some of the chapters here were presented at that session and subsequently updated in light of recent developments within the Basin. Several chapters are new additions coming from leading experts on Nile issues. The chapters from the Special Issue are either entirely or partially revised to reflect recent developments.

The chapters examine the implications of the GERD for the transboundary relations in the Nile Basin from law, politics, economics and hydrology disciplinary lenses, and focus on themes such as the need for cooperation and basin-wide approaches, the emergence of a change or a new governance regime, the role of fairness and equity, benefit-sharing, trade-offs, cooperation and coordination during dam filling and operation, and the wider issues around trade, investment and integration in the (Eastern) Nile Basin as a result of the GERD. The contributors to the volume are diverse in their backgrounds, disciplines and expertise, as well as their perspectives and conclusions. It is hoped that by drawing upon such a diverse range of contributors the volume will spark further debate, research and policy work on the challenges and the cooperative opportunities associated with the GERD.

The volume begins in the first chapter by articulating the key questions, arguments and findings of the collection. Chapters 2, 3 and 4 are dedicated to legal and theoretical questions and developments. The politics of the GERD and its basin-wide connotations are covered in Chapters 5 and 6. This is followed by Chapters 7 and 8 that look into the economic impacts of GERD; although they apply different methods, data and focus, their analysis and findings complement each other by providing similar, if not identical, insights into the economic impact of GERD. Chapters 9 and 10 offer a hydrological perspective on dam filling – applying different scientific methods, they offer alternative dam filling options to the stakeholders. While the first five substantive chapters are qualitative in nature, the other four are predominantly quantitative. The key message coming out of these chapters, and the combined experience and expertise of the contributors, is that if the riparian states can take full advantage of the opportunities provided for all by the GERD project, all will benefit.

The editors would like to thank those who have dedicated time and energy to peer-review earlier versions of the chapters. They are grateful to the IWRA, WI and Professor James Nickum for all the support that has been given. They are also thankful to all the contributors for their hard work, collegiality and collaborative spirit.

Zeray Yihdego, Alistair Rieu-Clarke, Ana Elisa Cascão
18 May 2017

References

Budapest Water Summit (2016). Messages and Policy Recommendations. Retrieved 15 May 2017, from www.budapestwatersummit.hu/data/images/Kepek_2016/BWS2016_Messages_Policy_Recommendations.pdf.

Sustainable Development Goals (2015). Retrieved 15 May 2017, from www.un.org/sustainabledevelopment/sustainable-development-goals/.

Abbreviations

AHD	Aswan High Dam
BCM	Billion cubic metres
CFA	Cooperative Framework Agreement
CGE	Computable General Equilibrium Model
CRU	Climatic Research Unit (University of East Anglia)
DoPs	Declaration of Principles on the GERD
DRC	Democratic Republic of Congo
DSS	Decision support systems
EAPP	East African Power Pool
EEPCo	Ethiopian Electric Power Corporation
ENA	Ethiopian News Agency
ENCOM	Eastern Nile Council of Ministers
ENSAP	Eastern Nile Subsidiary Action Program
ENTRO	Eastern Nile Technical Regional Office
ESIA	Environmental and Social Impact Assessment
FAO	Food and Agriculture Organization
GDP	Gross Domestic Product
GERD	The Grand Ethiopian Renaissance Dam
GTAP	Global Trade Analysis Project
GTP I	Growth and Transformation Plan (2010/2011–2014/2015)
GTP II	Growth and Transformation Plan II (of the Federal Democratic Republic of Ethiopia, 2015/2016–2019/2020)
HAD	High Aswan Dam
HASF	historical average streamflow
IBMR	Indus Basin Model Revised
ICJ	International Court of Justice
ICWC	International Centre for Water Cooperation (a UNESCO CAT II Centre)
IFAD	International Fund for Agricultural Development of the United Nations
IFPRI	International Food Policy Research Institute
ILC	International Law Commission
ILO	International Labour Organisation

IPoE	International Panel of Experts on the GERD
IWRA	International Water Resources Association
JMP	Joint Multipurpose Project
kWh	kilowatt-hour
LDCs	Least Developed Countries
MDI	consulting engineers
MFA	Ministry of Foreign Affairs
MIT	Massachusetts Institute of Technology
MoFED	Ministry of Finance and Economic Development, Ethiopia
MW	Megawatt
MWRI	Management of Water Resources and Irrigation
NBI	Nile Basin Initiative
NELSAP-CU	Nile Equatorial Lakes Subsidiary Action Program Coordination Unit
Nile-COM	Council of Ministers of Water Resources of the Nile Basin Countries
PCIJ	Permanent Court of Justice
RAPSO	Reservoir and Power Station Operation.
RWSM-Pak	Regional Water System Model for Pakistan
SAPs	Subsidiary Action Programmes
SDDP	Stochastic Dual Dynamic Programming
SDGs	Sustainable Development Goals 2015
SVP	Shared Vision Program
TECCONILE	Technical Cooperation Committee for the Promotion of Development and Environmental Protection of the Nile Basin
TFP	total factor of production
TNC	Tripartite National Committee; also Technical National Committee[1]
TWH	terawatt-hour
UN	United Nations
UN CESCR	United Nations Committee on International Covenant on Economic, Social and Cultural Rights
UNDP	United Nations Development Programme
UNECE	United Nations Economic Commission for Europe Convention
UNECE WC	Convention on the Protection and Use of Transboundary Watercourses and International Lakes 1992
UNESCO	United Nations Educational, Scientific and Cultural Organization
UNWC	Convention on the Law of the Non-Navigational Uses of International Watercourses 1997
USBR	United States Bureau of Reclamation
VCLT	Vienna Convention on the Law of the Treaties
WI	*Water International*

Note

1 Some reports and an English translation of the DoPs 2015 refer interchangeably to the Tripartite National Committee (TNC) as the Technical National Committee (TNC), which is explained by the technical nature of the Committee, however, TNC is officially the abbreviation for Tripartite National Committee.

1 A multi-disciplinary analysis of the risks and opportunities of the Grand Ethiopian Renaissance Dam for wider cooperation in the Nile

Zeray Yihdego, Alistair Rieu-Clarke and Ana Elisa Cascão

Introduction

The Nile River is the longest river in the world, travelling for 6,695 kilometres from its source in the Great Lakes region until its discharge into the Mediterranean Sea in Egypt. The river passes through eleven riparian countries: Tanzania, Uganda, Rwanda, Burundi, the Democratic Republic of the Congo, Kenya, Ethiopia, Eritrea, South Sudan, Sudan and Egypt. The main sources or sub-basins of the Nile are the Blue Nile, the White Nile and Tekezze/Atbara. While the White Nile is the longest sub-basin of the Nile, it only makes an estimated 15 per cent steady water contribution to the Nile; in contrast, the Blue Nile which originates in Ethiopia makes 'up to 90% of annual Nile flows' (NBI 2012: 26). However, this annual flow figure masks the significant variability in seasonal flows.

Hundreds of millions of people live in the Nile Basin. With respect to the former, the Nile Basin Initiative Atlas (NBI 2016) highlights that:

> The current total population of Nile Basin countries is estimated at 487.3 million. Ethiopia has the highest population (99.4 million) closely followed by Egypt (91.5 million) and DR Congo (72.1 million). Eritrea (5.2 million), Burundi (11.2 million) and Rwanda (11.7 million) have the smallest populations.
>
> (NBI 2016: 53)

The Atlas further provides that:

> the combined population living within the Nile Basin (covering an area of 3,176,541 km²) is estimated at 257 million (or 53% of the total population of Nile Basin countries). Egypt has the highest population living within the Nile Basin (85.8 million), followed by Uganda (33.6 million), Ethiopia (37.6 million) and Sudan (31.4 million). Eritrea (2.2 million) and DR Congo (2.9 million) have the smallest populations within the Nile Basin.
>
> (NBI 2016: 15)

The total population of Nile Basin countries will drastically increase in the next three or so decades. It is reported that,

> by the year 2050, annual increases (in African population) will exceed 42 million people per year and total population will have doubled to 2.4 billion. This comes to 3.5 million more people per month, or 80 additional people per minute.
>
> (Bish 2016)

Population dynamics are therefore set to have a real impact on the water resources management, development and allocation among Nile riparian countries.

The Grand Ethiopian Renaissance Dam (GERD) is a large-scale hydropower dam project, which has been under construction on the Blue Nile since 2011 – although the construction contract between the contractor, Salini Impregilo, and the Ethiopian Electric Power Corporation, was signed by the Ethiopian government in December 2010. The GERD is located in the Ethiopian portion of the Blue Nile Basin, close to the Sudanese border. With a dam height of 155 metres, 1,800 metres long, with a storage capacity of 74 CBM and with an upgraded installed capacity of 6,450 MW in electricity, the GERD will, on completion, be the largest dam on the African continent (Feyissa 2017). Ethiopia has recently announced that more than 57 per cent of the construction of this US$4.7 billion dam project has been completed (Feyissa 2017).

With an expected completion date of 2017, it is relevant and timely to explore the GERD's regional significance to the Nile Basin generally and more specifically to the Eastern Nile sub-basins. The need for the Eastern Nile riparians (Ethiopia, Egypt, Sudan and South Sudan) to reconcile their current and future interests with respect to Nile water resources in an equitable and sustainable manner is evident, yet challenging, especially given their growing populations.

A further challenge is that the countries of the Nile Basin are both underdeveloped and diverse in terms of their socio-economic conditions (NBI 2012). The 2016 UNDP Human Development Report, for instance, classifies the majority of Nile Basin countries in the category of 'low human development', while Egypt and Kenya are classified in the 'medium human development' group. The importance of geography to socio-economic conditions is highlighted in the Nile Basin Report (2012), which suggests that,

> [Egypt] has the advantage that most of its population live in the narrow tract of land along the Nile and in the Nile Delta areas, and its economy benefits from oil revenues. The headwater countries, in particular Tanzania, Kenya, Uganda, and Ethiopia, have been constrained in their efforts to provide similar quality of life for upstream riparian communities by the scattered settlement patterns and difficult terrain in the headwater areas.
>
> (Nile Basin Report 2012: 108)

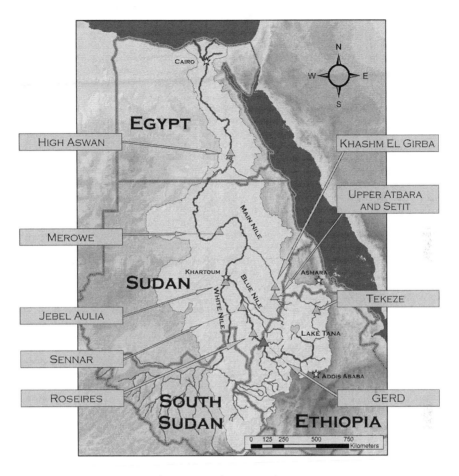

Figure 1.1 Map of Eastern Nile region, with reservoir locations.
Source: Wheeler *et al.* 2016: 613.

Alleviating poverty within these countries requires improved food and energy security for all. Rural populations in both Ethiopia and Sudan have extremely low levels of access to renewable energy (2 per cent and 7 per cent, respectively). Yet hydropower potential in the basin is high, particularly in the Ethiopian highlands where the hydropower potential is estimated at 45,000 mw (Salman, Chapter 2; GTP II 2015). At the same time, there is also significant potential to increase intra-basin power trade, between Egypt, Ethiopia, Sudan and beyond the Eastern Nile sub-basin (Tan *et al.* 2017). However, any development is heavily contingent upon the equitable and inclusive sustainable development of the waters of the Nile River Basin. Moreover, food and energy security are intricately linked. Any development in irrigation to help address the pervasive undernourishment will have to be reconciled with potentially competing interests such as hydropower. While reconciling these potentially competing interests poses significant challenges within sovereign borders, Nile Basin states must also account for the needs and interests of other riparian states.

Addressing the multiple socio-economic challenges faced by Nile Basin states demands a regional approach. An intergovernmental platform, the Nile Basin Initiative (NBI), was established in 1999 as a transitional arrangement gathering together all the Nile Basin riparians. Since its establishment, the NBI has been working towards promoting and advancing regional approaches to the social, economic and environmental challenges faced by the basin. However, population growth, the urgent need for sustainable development and the slow pace of cooperative efforts are placing more and more pressure on governments to move ahead with projects, even if they are mainly national in nature. As originally conceived, the GERD is one such example – although it is certainly not the first project of its kind in the Nile Basin. An assessment of both the Blue and White Nile Basins demonstrates that, for example, Egypt, Sudan and Uganda have all unilaterally developed large-scale infrastructure projects, particularly over the last two decades (Salman, Chapter 3). The GERD, however, is arguably the most significant given its size, location, its influence on transboundary relations between three key Eastern Nile Basin riparians, and its potential effect on regional hydropolitical relations at the Nile Basin level.

The GERD offers both risks and opportunities. As an opportunity, the GERD has the potential to foster cooperation by offering regional socio-economic benefits through the coordination and management of hydraulic infrastructures within the basin. Such benefits, it is hoped, will eventually contribute to an improved and a more efficient water management regime. An improved and more efficient water management regime may in turn assist in addressing the uncertainties that climate change is expected to bring to the basin – although it is not yet certain if this will mean more or less availability of resources (Siam and Eltahir 2017; Conway 2017). Coordination over the operation of the GERD and the other existing infrastructures may also prove to be a catalyst for additional benefits 'beyond water', such as a greater integration of markets and trade (Sadoff and Grey 2002), including in the energy sector. However, one of the most notable challenges in realizing these benefits is to ascertain and gain a broader agreement on the most appropriate legal, political and institutional arrangements that should be put in place among the concerned states.

Reaching that agreement has proven to be a significant challenge. In particular, because downstream and upstream riparians have historically displayed major differences in their perceived entitlements to the Nile waters. While downstream Egypt and Sudan have relied upon what they consider to be their 'historic rights', upstream riparian countries reject such claims and considered it to be contrary to the principle of equitable and reasonable utilisation and participation (Salman, Chapter 3).

It is not surprising then that, at least up until March 2015, the GERD was seen, particularly through the eyes of the media, as a source of political tension between Ethiopia and Egypt (Tawfik and Dombrowsky, Chapter 6). Egypt considered it as a violation of the 1902 Anglo-Ethiopian Nile Treaty, the 1993 Framework Cooperation instrument signed between Egypt and Ethiopia, and established principles of international water law – namely, the duty to take all appropriate measures to prevent significant harm and the obligation to protect the ecosystems of an international watercourse, in particular (Egyptian Ministry of Foreign Affairs 2014). Ethiopia for its part maintained its long-standing position on the invalid nature of old Nile agreements, including the 1929 and 1959 Nile Water Agreements, and based its actions in furtherance of the GERD project on its right to make use of Nile waters. In maintaining its right to use the Nile waters, Ethiopia placed considerable emphasis on the fact that it contributes the largest share to the total Nile flows. Ethiopia further argued that the GERD would bring shared socio-economic and environmental benefits for all the three riparian countries, rather than causing harm to the two downstream countries (Horn Affairs 2014).

The environmental and economic implications of the dam in all its aspects (e.g. hydropower, agriculture, etc.) on Sudan and Egypt is open for debate. Egypt, in particular, has concerns over the short-term and long-term impacts of the dam on Egypt's power generation capacity, irrigation potential and its economy in general. Ethiopia, for its part, insists that the hydropower project will bring economic, environmental and regional benefits even beyond the three Eastern Nile countries. Understanding the interests and concerns of the three countries with respect to the filling and operation of the GERD entails both challenges and opportunities, as discussed at length by Boehlert, Strzepek and Robinson (Chapter 7), Kahsay, Kuik, Brouwer and Van der Zaag (Chapter 8), Zhang, Erkyihun and Block (Chapter 9) and Wheeler (Chapter 10) in this book.

Despite the inevitable rounds of negotiations, the riparian states, assisted by expert studies, including an International Panel of Experts, have managed to navigate a cooperative path that seeks to reconcile their different interests (IPoE 2013; DoPs 2015; The Khartoum Document, December 2015, as referenced by Salman [Chapter 3]). The willingness of the riparian states to cooperate is reflected in the Declaration of Principles (DoPs) of March 2015, which endorsed established principles of international water law, cooperative mechanisms, and set out an agreement on how and which benefits of the dam will be shared and any negative impacts prevented. Although rather in a slow pace, the three Eastern Nile Basin countries are also undertaking additional environmental and economic impact studies on the GERD, as recommended by the IPoE 2013, through two French consultancy firms, BRLi Group and Artelia. It ought to be stressed that Sudan played a pivotal role in the trilateral negotiations on the GERD from the very beginning, and provided official backing to the project while constantly highlighting the downstream benefits of the GERD (see Cascão and Nicol, Chapter 5; Salman, Chapter 3).

Two observations recap the broader context of the GERD within the Nile Basin. First, although the attention of scholars is often focused on the views of, and relations between, Egypt and Ethiopia, the project is of great importance to the whole Nile Basin region including the multilateral cooperation processes within (Cascão and Nicol, Chapter 5). Second, while notable successes with respect to the multilateral cooperation processes have been achieved (see also technical cooperation under the NBI and political negotiations under the Cooperative Framework Agreement [CFA]) (Salman, Chapter 2), reaching a permanent legal and institutional framework that is accepted by all co-riparians remains a key challenge (Cascão and Nicol, Chapter 5).

Chapter overview of the volume

The contributions in this volume explore – from a range of disciplinary lenses – challenges and opportunities associated with GERD. One such perspective is law, which is considered from different angles in the first three chapters of the volume.

Following the current introductory chapter, Salman's Chapter 2 considers the ways in which the 'Gordian Knot' associated with the CFA can be disentangled. The chapter therefore provides an important link between basin wide cooperative efforts and the specific context of the GERD. After providing a geographical, political and historical background to the broader question of sharing Nile waters, Salman outlines the controversies relating to colonial and post-colonial Nile treaties and agreements. This then leads to an in-depth exploration of the attempts made to foster basin-wide cooperation through, in particular, the CFA. Salman highlights that, although the main objective of the NBI since 1999 was the conclusion of 'a cooperative framework agreement', the major disagreement over historic or acquired versus equitable rights continued to be divisive in negotiations up until the adoption of the CFA. The CFA, Salman notes, was, while not without its controversies, an important step change in the evolution of international water law within the Nile Basin. Taking into account new realities and mediation efforts, including the DoP, Salman goes on to maintain that the events related to the GERD have the potential to disentangle the Gordian Knot of the CFA.

In Chapter 3, Salman explores the contents of the DoPs on the GERD and the processes thereof, and asks whether or not this development levels the Nile Basin's playing field. In so doing, the chapter builds upon Salman's previous chapter. As a background, Salman begins by looking at the relevant legal instruments and the history of dams in the Nile Basin. He provides a detailed account of the sequence of negotiations that led to the DoPs and the December 2015 Khartoum Document, which endorsed the decision to have French firms BRLi and Artelia conduct an impact study on the GERD. Lamenting the prevalence of unilateral dam development in the Nile Basin and highlighting the absence of water security of the two downstream countries in the Declaration, Salman argues that the DoPs and the December Document have brought 'a new legal order' that has replaced the 1902 and 1959 Nile treaties – one that is founded upon contemporary principles of international law. He further argues that through the endorsement of the fundamental principles of international law, 'the playing field of the Nile Basin have been levelled'. Such levelling of the playing field should, he maintains, be seized upon as an opportunity to promote cooperation and collective action for sustainable and optimal utilisation of Nile waters, which will secure the interests of current and future generations.

In Chapter 4, Yihdego and Rieu-Clarke examine how the fairness principle – as articulated in Thomas Franck's fairness discourse (1995) – helps explain the strengths and weaknesses of the Nile legal framework both in terms of distributive equity and 'right process'. The chapter submits, first, that, while certain improvements could be made, the principle of fairness is sufficiently imbedded into international water law in light of the well-established principle equitable and utilisation, and its call for distributive justice, and in the many procedural requirements such as the duty to notify and consult, which enhance the legitimacy of any treaty regime. Second, it suggests that the way in which old Nile treaties came into being and their substance fail to satisfy a fairness test for a number of reasons. In contrast, the authors cautiously suggest that post-1990 Nile Basin initiatives and legal frameworks are more aligned with Franck's notion of fairness, and, while they also have shortcomings, are therefore likely to be complied with. This is particularly true with the 2015 DoP. Third, Yihdego and Rieu-Clarke maintain that a basin or regional approach to cooperation, particularly if it engages with non-state actors at multiple levels, might help rectify imbalances in power among riparians and ensure fairness over arbitrariness in the (Eastern) Nile Basin. Finally, the authors, based upon the effort they have made to look into the realization and determination of distributive justice using quantitative scholarship, found that an interdisciplinary approach to the fairness question may well delineate the application of the concept and ultimately bolster compliance with legal engagements involving the sharing of natural resources between states.

The law-related contributions are then followed by two other chapters that employ a political science perspective to the GERD – one focusing on hydropolitical developments that preceded the GERD and their implications for the cooperation processes, and the second adopting a broader approach to identifying the political origins and the future of the GERD.

Chapter 5 by Cascão and Nicol provides a comprehensive critical assessment of the GERD as both an outcome of change and as a catalyst for future change. The authors begin with a background discussion on the cooperation processes prior to the announcement of the GERD, both at the basin and sub-basin (Eastern Nile) level. They provide a detailed analysis of the achievements, as well as the pitfalls and challenges, faced in both cooperative tracks – the technical (NBI) and political (Cooperative Framework Agreement, CFA). The authors argue that the GERD and related norms and processes are partly the outcome of the significant changes in the transboundary relations that have been taking place since the mid-1990s. They also analyse the GERD as a shaper of future cooperation – in terms of its capacity to provide opportunities to enhance shared economic benefits and trade in the field of energy; and as an opportunity to expand regional development and integration in the Eastern Nile Basin region. The lessons learnt from the GERD are: first, cooperation at the sub-basin and basin-wide forums can coexist, but in the long term, might be hindered by high transaction costs associated with a plethora of concomitant processes, and second, that the GERD provides a strategic opportunity on many levels, including economic, political and diplomatic, to highlight the pressing need for a basin-wide transboundary water regime.

In Chapter 6, Tawfik and Dombrowsky discuss whether the developments associated with the GERD, and the DoPs in particular, transformed the hydro-diplomatic relations in the eastern Nile Basin countries from a water-sharing to benefit-sharing paradigm. The authors suggest that the dam could bring benefits to all riparians – economic development (in the form of power generation and power trade) for Ethiopia, more flow of water and electric generation to Sudan, and evaporation reduction from the High Aswan Dam and electricity provision to Egypt and Sudan. Furthermore, they suggest that the dam could be a catalyst for regional integration through power trade within and outside the eastern Nile Basin. Nonetheless, the authors caution that the GERD's downstream impacts, in particular to Egypt, will rest on dam filling and operating strategies. In consideration of the impacts, the authors raise important questions about the impacts of the dam at local, national and transboundary levels, including the issues surrounding its finances, climate change, as well as environmental and social impacts. Given that it is a unilateral project financed, owned and managed by one of the parties, the authors submit that the GERD project has not yet progressed from water-sharing to benefit-sharing. However, the authors conclude that 'the GERD is changing hydropolitics relations on the eastern Nile'. The authors contend that whether that change will lead to cooperation or non-cooperation – due to a lack of incentives for cooperation from the Ethiopian side because of sustained economic growth and the country's growing political power in the region – remains to be seen.

The two subsequent chapters, Chapters 7 and 8, provide an economics perspective on the implications of the GERD for both downstream and upstream Eastern Nile Basin countries, and examine the trade-offs that can be achieved to enhance cooperation in the basin.

In Chapter 7, Boehlert, Strzepek and Robinson examine the economy-wide impacts of the GERD filling options on Egypt. The authors begin by setting out three models for their economic analysis: a computable general equilibrium (CGE) model of the Egyptian economy; a water systems model of the Nile upstream of Lake Nasser which encompasses the GERD; and a water management model of the Egyptian water system. Applying a risk-based approach to take account of the unknowns of future water flows into the Nile, they analyse three dam filling scenarios to understand the economic impacts of the GERD to Egypt (and Ethiopia). The first one is 'the unconstrained scenario', which assumes that the dam will be filled swiftly, depending on available water flow from the Blue Nile. The second one envisages a three-year filling period; and the third one a ten-year filling period. Each of these scenarios considers three release requirements while the GERD is filling: no minimum release, and a 15 BCM and 30 BCM minimum release each year. Based upon such methods and approaches, they examine the impact of the GERD filling options on Egyptian hydropower, agricultural deliveries, in effect on Egypt's gross domestic product (GDP) during dam filling. The authors conclude that the worst-case impacts of unconstrained filling of the GERD on Egypt's economy are modest, and that Egypt's economy would not therefore be significantly affected by the filling of the GERD regardless of the option that is chosen. Additionally, the authors maintain that the higher short-term gains to the Ethiopian economy would mean that a more rapid filling policy would have higher Nile-wide economic benefits. Such conclusions are not withstanding the need for further research on expanding the method used in the chapter to understand the trade-offs between the Sudanese and Ethiopian economies and water uses relating to the GERD.

In Chapter 8, Kahsay, Kuik, Brouwer and Van der Zaag, as a complement to Chapter 7, look at the impacts of the dam during impounding and operational phases, the basin-wide economic implications, including to Ethiopia's economy and the impacts of the dam to the wider economy of the three Eastern Nile Basin countries. Unlike Boehlert, Strzepek and Robinson, who use a single country Computable General Equilibrium (CGE) model, the authors employ a multi-region multi-sector analysis to explore the GERD's basin-wide economic impacts in a global CGE setting. This is supported by the partial equilibrium analysis, which uses detailed information on specific economic sectors and projects such as hydropower plants. Once the GERD becomes operational, the authors' study identifies wide-ranging benefits and potential negative impacts associated with the project. The authors maintain that there are substantial basin-wide economic benefits from the GERD. However, the authors adopt a more cautious approach by suggesting that in order to avoid adverse effects on the Egyptian economy when the dam is filled, it would be advantageous to extend the impounding period of the dam if it occurs during dry years, so as to mitigate any significant negative impacts. The authors go on to suggest that, given the susceptibility of Egyptian hydropower to adverse impacts as a result of the GERD filling and operation phases, and to strengthen the basin-wide significance of the dam, there is a need for the countries to institute a basin-wide power trade scheme that allows Egypt to buy electricity from the GERD. The authors also recognise a need for further study on issues such as 'the potential impact of climate change on the development of the GERD and hence its transboundary economic impacts'.

While adopting differing methods and scope, the last two chapters bring a complementary hydrological perspective to the analysis of the GERD.

In Chapter 9, Zhang, Erkyihun and Block address the critical issues of initial dam management: characterising inflow and reservoir filling strategies. The authors capture inter-annual and multi-decadal streamflow characteristics using a wavelet analysis to model expected streamflow conditions during reservoir filling. This allows for an improved understanding of the likely shifts towards wetter or drier conditions during this critical phase. Worthy of note here is that similar, and at times identical considerations, have been used by both Kahsay *et al.* (Chapter 8) and Boehlert *et al.* (Chapter 7), albeit from an economics perspective. Zhang *et al.* analyse various reservoir filling strategies and associated impacts on upstream and downstream countries, and consider questions over who bears the risks associated with natural streamflow variability. Three strategies for filling the GERD are considered: impounding a pre-defined fraction of streamflow each month (e.g. 10 per cent, 25 per cent); only impounding streamflow when conditions are wetter than average; and filling within a pre-defined number of years (e.g. four, six, eight years). To compare outcomes, simulations of expected streamflow are generated using the wavelet analysis and subjected to each of the filling strategies. The authors maintain that the four-year filling strategy results in the most rapid filling, but it is likely to induce sharp reductions in streamflow to Sudan and Egypt over that period. Conversely, the 10 per cent strategy requires more than ten years to fill, on average, limiting hydropower-generating ability. The 25 per cent, six-year and eight-year strategies appear to be compromise solutions, with filling and hydropower generation occurring at moderate rates. Yet all three strategies, the authors suggest, still express uncertainty in outcomes given climate variability. The authors call for closer cooperation and coordination among the three major riparian countries, not only regarding reservoir filling, but also with respect to the long-term management for purposes of fostering development and regional integration.

In Chapter 10, Wheeler presents a new 'hydro-policy' modelling framework to simulate complex multi-objective reservoir operations throughout the Eastern Nile Basin. Developed using the RiverWare modelling platform, the aim of the framework is to provide stakeholders with a tool that can accurately simulate the operational decisions of a managed river system, encourage understanding of how the model functions through its transparent rule-based logic and architecture, and be highly flexible to allow its use as an analytical tool to support negotiations between stakeholders. Wheeler demonstrates the framework's application by analysing potential coordination and adaptation strategies among the dams of Ethiopia, Sudan and Egypt during the initial filling of the GERD reservoir. Through this analysis, the author suggests that the time required to fill the GERD is highly dependent on hydrologic conditions and releases made during this period. He cautions that while non-cooperative filling can occur in as little as three years, the risks to downstream countries under such a schedule is high. Additionally, Wheeler maintains that the downstream effects of filling the GERD will depend on both the agreed annual release and the storage in Lake Nasser at the time when filling begins. While the probability of reductions to Egypt's water availability ranges widely, downstream concerns can be substantially reduced by considering an annual agreed release of 35 BCM or greater throughout the filling period. Major consumptive uses in Sudan can be maintained by maximising the storage of the reservoirs throughout this period. Most significantly, the author contends that the risks of 'unplanned shortages' can be significantly reduced or eliminated during the filling phase of the GERD by both invoking the drought management plan of the High Aswan Dam and with a collaborative 'safeguard' policy that allows additional releases from the GERD under critical circumstances in Lake Nasser. Ultimately, the author concludes that while increased risks to downstream users certainly do exist during the filling period, they can only be minimised with an explicit coordination plan. A basic level of coordination with an agreed annual release substantially, but only partially, reduces risks to Egypt. A more robust strategy is continuous coordination throughout the filling process by considering the conditions of the High Aswan Dam. The author also notes that any cooperative solution must be built on transparent sharing of information.

Findings and insights

This volume systematically examines the question of whether the GERD is contributing to shape the changing dynamics in the Nile Basin through four disciplinary lenses – law, politics, economics and hydrology. These different but complementary perspectives provide both a qualitative and quantitative account of the implications of the GERD for basin-wide cooperation. The insights from each of these diverse chapters are too rich to cover here comprehensively. Eight major themes related to the risks and opportunities of the GERD for promoting cooperation will therefore be offered.

First, it appears that the GERD has brought a new era of cooperation in the (Eastern) Nile Basin; the different phrases used by contributors, for example, 'new legal order' by Salman, a game or power changer by Tawfik and Dombrowsky, and Cascão and Nicol, and a 'fair system' by Yihdego and Rieu-Clarke make this point abundantly clear. The negotiating process and the deal struck between the three Eastern Nile riparians countries is fundamentally different from the previous regime that existed. This may, however, be challenged from two angles. One is that the new deal has neither expressly rejected nor endorsed the rules of the 'old Nile regime' (i.e. one characterised by the 1959 Agreement). Hence, as Tawfik and Dombrowsky argue, Egypt remains concerned that the DoPs does not guarantee the 'historic share' of downstream countries. Against this, Salman, and Yihdego and Rieu-Clarke, see the new system of cooperation as a fresh beginning, which rejects perceived entitlements based upon old treaties. The other angle is that the GERD dynamics are not only an outcome of post-2011 negotiations, they can also be traced back to the multilateral basin-wide initiatives that resulted in the establishment of the NBI and the signing of the CFA (Cascão and Nicol, Chapter 5). In this regard, the deadlock in cooperation between downstream and upstream countries surrounding the CFA may well have been unlocked by the GERD.

Second, the recognition of the GERD as having beneficial outcomes regionally and nationally, in terms of energy production and trade, sustainable utilisation of the water resource and potential socio-economic benefits, is worthy of stressing. Despite the uncertainties and disagreements over the degree of adverse impacts that the GERD might have in the short-term, and the level of potential benefits, the contributions in this volume suggest that the concerned countries will gain from the GERD (a theme separately treated later). The national, sub-basin and basin-wide economic benefits are well recognised as expressly articulated in quantitative terms by Boehlert *et al.* and Kahsay *et al.* and in qualitative terms by Tawfik and Dombrowsky and Salman (Chapter 3). The potential and emerging trade, investment and energy sharing mechanisms and plans at the sub-basin level (and beyond) might be an important positive first step towards wider benefit-sharing approaches among neighbouring riparians and the expansion of the regional integration process. For example, in response to an Egyptian interest to purchase electric power from Ethiopia, Ethiopia has recently revealed that it is finalising 'a feasibility study to export electricity to Egypt' (ENA 2017). However, as stated by Wheeler, Zhang *et al.*, and Tawfik and Dombrowsky, concerns over hydrological, environmental and other downstream impacts of the dam do remain. The ongoing endeavours of the parties to address their concerns, with the backing of national, regional and international expert studies and foreign consultancy firms, are commendable, despite the complexity of these studies, talks and processes. It is argued that due to the GERD's unilateral nature, including being financed and managed by one riparian and the many unresolved concerns to downstream hydropower and other water uses, the GERD's benefit-sharing outcome is certainly not guaranteed (Tawfik and Dombrowsky, Chapter 6). As this volume demonstrates, the potential for multi-legal benefit-sharing is evident (Boehlert *et al.*, Chapter 7; Kahsay *et al.*, Chapter 8). However, this will require tackling the challenges and concerns of all the parties involved, and identifying and maximising available incentives. Finding a common ground to mitigate financial and technical burdens related to the GERD will help the parties to be fully engaged in a negotiated settlement of outstanding issues. Building upon already existing joint initiatives such as the NBI and its associated programmes offers the potential for further cooperation and integration in the Nile Basin (Cascão and Nicol, Chapter 5; Tawfik and Dombrowsky, Chapter 6; Kahsay *et al.*, Chapter 8; Zhang *et al.*, Chapter 9).

Third, in the short and medium terms, one of the most important issues pertains to the filling of the reservoir of the GERD. Decisions taken in that regard are extremely important to an understanding of how the GERD might actually influence a 'new and fair legal order', or constitute a 'game-changer' or foster 'benefit-sharing' and regional integration. The two hydrology chapters by Zhang *et al.* and Wheeler provide some solutions to the question of filling, with the former providing a more the GERD-focused solution to filling, and the latter also accounting for potential downstream infrastructure. Both chapters emphasise that a negotiated compromise needs to be established to minimise harm on Sudan and Egypt during filling (Zhang *et al.*, Chapter 9; and Wheeler, Chapter 10). Similarly, the two economics chapters have dealt with the question of the economic impacts of the GERD filling, although Boehlert *et al.* focuses on the impact of filling on the Egyptian economy during filling, Kahsay *et al.* looks at the broader economic implications to the three Eastern Nile countries and their wider economy beyond agriculture and hydropower. Both chapters offer a variety of possible solutions to mitigate the short-term effects of the dam filling and to maximise the economic benefits of the project beyond water sharing. Worthy of note is that while Boehlert *et al.* propose a rapid filling as being more beneficial to the economy of the three countries at stake and the Nile Basin more generally, Kahsay *et al.* suggest a reasonably longer filling as a solution to reducing the risks of filling to Egypt.

Fourth, although the two hydrology chapters have not dealt with post-filling operation of the dam, the authors have reiterated that a coordinated and mutually beneficial dam operational management scheme would be a key to effective cooperation and integration in the basin (Zhang *et al.*, Chapter 9). The importance of coordinating the operations of the GERD, the High Aswan Dam and the multiple Sudanese reservoirs, in particular during drought situations has been stressed (Wheeler, Chapter 10). The need for cooperation and promoting shared economic benefits for all has also been considered from an economics perspective.

Fifth, the GERD's basin-wide implications should not be understated. As the principles adopted in the DoPs are consistent with key parts of the CFA, and the overall political vision and ambition of the NBI, the three countries may well be encouraged, depending of course on many national and regional factors, to elevate or resume their cooperation at a basin-wide level. This was seen by Salman, Yihdego and Rieu-Clarke, and Tawfik and Dombrowsky as a particularly appropriate remedy to the ongoing practice of unilateralism in the Nile Basin. Other Nile countries might be encouraged to sign and ratify the CFA. Noteworthy is the fact that these processes operate in parallel. Tanzania, for example, ratified the CFA a few days after the DoPs were signed by Ethiopia, Egypt and Sudan. In the long run, the negotiation processes and trust-building exercises related to the GERD may be an important stepping stone in addressing some of the sticking points that Nile Basin countries faced when negotiating the CFA (Salman, Chapter 2; and Tawfik and Dombrowsky, Chapter 6).

While the focus of the volume is not the CFA *per se*, the volume demonstrates that states promote their interests across international and regional cooperative platforms based upon established norms and processes. As highlighted by Cascão and Nicol, and Tawfik and Dombrowsky, Egypt, Ethiopia and Sudan are participating in the GERD trilateral process; and the last two are taking part in NBI cooperation processes, while Egypt may (or may not) return as a full-member to the NBI – which is currently under discussion at the highest political levels. As with the synergies between global and basin-level process, these regional efforts offer opportunities to foster the 'cross-fertilisation' of established norms across a range of different cooperative frameworks; and also offer different venues by which to develop a shared understanding of the key legal rules and principles. In this regard, and given the challenges of achieving a legal framework in the near future, the NBI should be strengthened as a good avenue for cooperation in the Nile Basin (Tawfik and Dombrowsky, Chapter 6).

Sixth, the outstanding issues to be resolved regarding the GERD (and the Nile Basin more generally) must be based upon the readiness of all concerned to make compromises. This was particularly stressed by Salman (Chapter 2), Tawfik and Dombrowsky, Boehlert *et al.* and Yihdego and Rieu-Clarke. Areas of compromises include managing potential impacts and trade-offs relating to the GERD, resolving the basin-wide endeavour to establish a permanent legal and institutional framework in accordance with the dictates of internationally recognised principles and procedures, including equity. In this regard, two points need to be emphasised: (a) the rejection of making unqualified claims, which is inimical to reaching a fair and equitable compromise (Yihdego and Rieu-Clarke, Chapter 4); and (b) the importance of identifying incentives to help the parties commit to a negotiated positive-sum solution with respect to the GERD and other potential projects (Tawfik and Dombrowsky, Chapter 6).

Seventh, the need for a transparent and participatory decision-making at project, national and trilateral levels has been emphasised by various chapters of the volume. Future negotiations and cooperation that are crucial to turn the terms of the DoPs and the Khartoum document into reality should therefore respond to these demands. Legitimate stakeholders including concerned riparians (Wheeler, Chapter 10), the NBI (Tawfik and Dombrowsky, Chapter 6), scholars, civil society organisations and citizens (Yihdego and Rieu-Clarke, Chapter 4; Tawfik and Dombrowsky, Chapter 6) must have a fair say in both the substance and processes of sharing and protecting (Blue) Nile waters. In the short and medium term, it has been specifically emphasised that the GERD negotiations should not be limited to the Tripartite National Committee (TNC) – regular and institutionalised ministerial meetings and other avenues must be used to broaden the scope and quality of negotiations (Tawfik and Dombrowsky, Chapter 6).

Finally, the signing of the DoPs, which relates to a complex sub-basin, and coincidentally followed the entry into force of the UN Watercourses Convention, evidences the interrelationship between global and regional developments. The DoP is not only consistent with the CFA but also with the UN Watercourses Convention, and this might mean that the three Eastern Nile and the other Nile Basin countries may well be encouraged to consider joining the UN Watercourses Convention, subject to appropriate study, debate and constitutional processes of each riparian. Now, also with the opening of the United Nations Economic Commission for Europe (UNECE) Convention on the Protection and Use of Transboundary Watercourses and International Lakes and a commitment under the Sustainable Development Goals related to integrated water resources management at the transboundary level, further opportunities are available to capitalise on the synergies between, and momentum behind, these global and basin-level processes.

By way of conclusion, this volume explores the opportunities and challenges surrounding the GERD through multiple lenses. The results and suggestions offered by the eighteen experts, with a wealth of experience, make a unique and well-rounded contribution to knowledge relating to the utilisation and preservation of Nile Basin water resources. It is hoped that the insights set forth in this volume shall inform policy both in the short and medium terms, with respect to the ongoing talks and studies related to the GERD, and in the longer term, efforts needed to foster and strengthen basin-wide cooperation, integration and economic development throughout the region. It is important to remind readers that the volume does not pretend to have answers to all questions arising from the GERD or the Nile Basin issue. For example, and as mentioned in one of the chapters, a study on the impacts of climate change in realizing the ambitions of the (Eastern) Nile Basin countries would be necessary and timely. To respond to the theme mentioned in the title – the volume clearly demonstrates there are multiple and far-reaching opportunities to grasp from the GERD project, if riparian states, supported by others, continue to follow a cooperative positive sum path.

References

Bish, J.J. (2016, 11 January). Population Growth in Africa: Grasping the Scale of the Challenge. *Guardian*.

Boehlert, B., Strzepek, K.M. and Robinson, S. (2018). Analysing the Economy-Wide Impacts on Egypt of Alternative GERD Filling Policies. In Z. Yihdego, A. Rieu-Clarke and A.E. Cascão (eds), *The Grand Ethiopian Renaissance Dam and the Nile Basin: Implications for Transboundary Water Cooperation*. London: Routledge.

Cascão, A.E. and Nicol, A. (2018). Changing cooperation dynamics in the Nile Basin and the role of the GERD. In Z. Yihdego, A. Rieu-Clarke and A.E. Cascão (eds), *The Grand Ethiopian Renaissance Dam and the Nile Basin: Implications for Transboundary Water Cooperation*. London: Routledge.

Conway, D. (2017, 24 April). Future Nile River Flows. *Nature Climate Change*. Published online.

DoPs (2015, 25 March). *Declaration of Principles between The Arab Republic of Egypt, the Federal Democratic Republic of Ethiopia and the Republic of the Sudan on the Grand Ethiopian Renaissance Dam Project* (GERDP). *Horn Affairs*. Retrieved 20 May 2016, from http://hornaffairs.com/en/2015/03/25/egypt-ethiopia-sudan-agreement-on-declaration-of-principles-full-text/.

Egyptian Ministry of Foreign Affairs (2014). Egypt's Perspective Towards the Ethiopian Grand Renaissance Dam Project (GERDP). Retrieved 21 May 2016, from www.mfa.gov.eg/English/EgyptianForeignPolicy/Pages/renaissance_dam.aspx.

ENA (2017). Ethiopia Finalizes Feasibility Study to Export Electricity to Egypt. 30 April. Retrieved 15 May 2017, from www.ena.gov.et/en/index.php/economy/item/3129-ethiopia-finalizes-feasibility-study-to-export-electricity-to-egypt.

Feyissa, G. (2017). Six Years of Thriving Journey on Construction of GERD'. ENA, 7 April. Retrieved 8 April 2017, from www.ena.gov.et/en/index.php/economy/item/3022-six-years-of-thriving-journey-on-construction-of-gerd-gosaye-feyissa-ena.

Franck, T. (1995). *Fairness in International Law and Institutions*. Oxford: Clarendon Press.

GTP, The Second Growth and Transformation Plan (GTP II), The Federal Democratic Republic of Ethiopia (2015). Retrieved 11 May 2017, from www.africaintelligence.com/c/dc/LOI/1415/GTP-II.pdf.

Horn Affairs (2014, 30 March). Ethiopia Snubs 'Egyptian Perspective' as Falsification and Populist. Retrieved 1 June 2016, from http://hornaffairs.com/en/2014/03/30/ethiopia-snubs-egyptian-perspective-falsification-populist.

International Panel of Experts (IPoE) (2013). Final Report on Grand Ethiopian Renaissance Dam Project (GERDP), 31 May. Retrieved 1 June 2016, from www.internationalrivers.org/files/attached-files/international_panel_of_experts_for_ethiopian_renaissance_dam-_final_report_1.pdf.

Kahsay, T.N., Kuik, O., Brouwer, R. and Van der Zaag, P. (2018). Economic Impact Assessment of the Grand Ethiopian Renaissance Dam. In Z. Yihdego, A. Rieu-Clarke and A.E. Cascão (eds), *The Grand Ethiopian Renaissance Dam and the Nile Basin: Implications for Transboundary Water Cooperation*. London: Routledge.

NBI (2012). *State of the River Nile Report*. Retrieved 1 June 2017, from www.nilebasin.org/nileis/system/files/Nile%20SoB%20Report%20Chapter%204%20-%20Population.pdf.

NBI (2016). The Nile Basin Water Resources Atlas. Retrieved 1 May 2017, from http://nileis.nilebasin.org/content/nile-basin-water-resources-atlas.

Nile Treaty (1959). Agreement between the Republic of the Sudan and the United Arab Republic for the Full Utilisation of the Nile Waters, Cairo, 8 November 1959 (entered into force 12 December 1959), *Treaty Series*, XCIII, 43.

Sadoff, C. and Grey, D. (2002). Beyond the River: The Benefits of Cooperation on International Rivers. *Water Policy*, 4, 389–403.

Salman, S.M.A. (2018a). The Nile Basin Cooperative Framework Agreement: Disentangling the Gordian Knot. In Z. Yihdego, A. Rieu-Clarke and A.E. Cascão (eds), *The Grand Ethiopian Renaissance Dam and the Nile Basin: Implications for Transboundary Water Cooperation*. London: Routledge.

Salman, S.M.A. (2018b). Agreement on Declaration of Principles on the GERD: Levelling the Nile Basin Playing Field. In Z. Yihdego, A. Rieu-Clarke and A.E. Cascão (eds), *The Grand Ethiopian Renaissance Dam and the Nile Basin: Implications for Transboundary Water Cooperation*. London: Routledge.

Siam, M.S. and Eltahir, E.A.B. (2017, 24 April). Climate Change Enhances Interannual Variability of the Nile River Flow. *Nature Climate Change*. Doi: http://dx.doi.org/10.1038/nclimate3273.

Tan, C.C., Erfani, T. and Erfani, R. (2017). Water for Energy and Food: A System Modelling Approach for Blue Nile River Basin. *Environments*, 4(15). Doi: 10.3390/environments4010015.

Tawfik, R. and Dombrowsky, I. (2018). GERD and Hydropolitics in the Eastern Nile: from Water-Sharing to Benefit-Sharing?. In Z. Yihdego, A. Rieu-Clarke and A.E. Cascão (eds), *The Grand Ethiopian Renaissance Dam and the Nile Basin: Implications for Transboundary Water Cooperation*. London: Routledge.

UNDP (2016). Human Development Report. Retrieved 1 May 2017, from www.undp.org/content/undp/en/home/librarypage/hdr/2016-human-development-report.html.

Wheeler, K.G. (2018). Managing Risks While Filling the Grand Ethiopian Renaissance Dam. In Z. Yihdego, A. Rieu-Clarke and A.E. Cascão (eds), *The Grand Ethiopian Renaissance Dam and the Nile Basin: Implications for Transboundary Water Cooperation*. London: Routledge.

Wheeler, K.G., Basheer, M., Mekonnen, Z.T., Eltoum, S.O., Mersha, A., Abdo, G.M., Zagona, E.A., Hall, J.W. and Dadson, S.J. (2016). Cooperative Filling Approaches for the Grand Ethiopian Renaissance Dam. *Water International*, 41(4), 611–634. Doi: 10.1080/02508060.2016.1177698.

Yihdego, Z. and Rieu-Clarke, A. (2018). International Law Developments on the Sharing of Blue Nile Waters: A Fairness Perspective. In Z. Yihdego, A. Rieu-Clarke and A.E. Cascão (eds), *The Grand Ethiopian Renaissance Dam and the Nile Basin: Implications for Transboundary Water Cooperation*. London: Routledge.

Zhang, Y., Erkyihun, S.T. and Block, P. (2018). From Projecting Hydroclimate Variability to Filling the GERD: Upstream Hydropower Generation and Downstream Releases. In Z. Yihdego, A. Rieu-Clarke and A.E. Cascão (eds), *The Grand Ethiopian Renaissance Dam and the Nile Basin: Implications for Transboundary Water Cooperation*. London: Routledge.

2 The Nile Basin Cooperative Framework Agreement

Disentangling the Gordian Knot

Salman M.A. Salman

Introduction: political geography of the Nile Basin

The Nile is the longest river in the world, stretching some 6,660 kilometres from its origins in Burundi and Rwanda to the Mediterranean Sea. It has a basin area of more than 3 million square kilometres, extending over eleven countries that share the river, and covering one-tenth of the African continent. Out of the more than 450 million people of the eleven riparian countries in 2016, about 250 million live or depend on the waters of the river, and the figure is expected to reach 300 million by 2030.

The Nile River consists of two distinct basins, the White Nile and the Blue Nile. The White Nile has its origins in the springs rising from the hills of Burundi and Rwanda. These springs combine to form the Kagera River, the largest watercourse to flow into Lake Victoria. The lake is the main source of the White Nile, and is shared by Tanzania (49 per cent), Uganda (45 per cent) and Kenya (6 per cent) (Salman 2001).

After exiting from Lake Victoria, the White Nile, which is called at that point the 'Victoria Nile', flows through Lake Kyoga, and then Lake Albert, and the river is called thereafter the 'Albert Nile'. Lake Albert is shared, similar to the Semliki River that feeds it, by Uganda and the Democratic Republic of Congo. The Albert Nile thereafter enters the Republic of South Sudan, where it is renamed Bahr el Jebel. It is joined in South Sudan from the west by Bahr el Ghazal, which originates in the watershed area between South Sudan and the Democratic Republic of Congo, and from the east by the Sobat River, which originates in Ethiopia.

After it re-emerges from the swamps of South Sudan, which are called the Sudd, the river takes its generic name of the White Nile. Evaporation and seepage consume almost two-thirds of the flow of the river in the Sudd swamps. The White Nile crosses the Republic of South Sudan into Sudan, and is joined in Khartoum by the Blue Nile that originates in Ethiopia. Prior to joining the White Nile, the Blue Nile itself is joined by two rivers – the Dinder and Rahad that also originate in Ethiopia (Collins 1996).

After their merger in Khartoum, the White Nile and the Blue Nile form the River Nile, which is joined thereafter by the Atbara River, the last tributary to join the River Nile. The Atbara River originates in Ethiopia, with one of its tributaries, the Setit, crossing from Eritrea. The river thereafter flows through northern Sudan before crossing into Egypt, and emptying whatever is left of its waters into the Mediterranean Sea (Collins 2002).

The stakes of the eleven Nile Basin states, their interests and contribution to, and uses of, the Nile waters, vary considerably. The stakes and interests of Egypt, Sudan, South Sudan and Ethiopia in the Nile are classified as very high; those of Uganda as high; those of Tanzania, Kenya, Burundi and Rwanda as moderate; while Eritrea and the Democratic Republic of Congo have low stakes in the river (Salman 2011a). Ethiopia contributes about 86 per cent of the total flow of the Nile waters, while the Equatorial Lakes provide the remaining 14 per cent. Yet, Egypt and Sudan use almost the entire flow of the river. Naturally, this is where the disputes over the Nile Basin begin; and they are interwoven in what is termed the Nile 'colonial treaties', although not all of these treaties were concluded during the colonial era.

The purpose of this chapter is to give an overview of the main treaties con-cluded on the Nile Basin, and to highlight the areas of differences and disputes on these treaties. The chapter then outlines the attempts of the Nile riparians to move towards cooperation, and discusses the main features of the Nile Basin Cooperative Framework Agreement (CFA) and the differences thereon. The chapter concludes by setting forth some ideas for resolving the current impasse over the CFA, and trying to disentangle the CFA Gordian Knot.

Colonial treaties and disputes over the Nile waters

A number of treaties on the Nile Basin were concluded during the last two cen-turies, and most of them are causes of disputes. One treaty that is the source of a major dispute between Ethiopia on the one hand, and Egypt and Sudan on the other, is the 'Treaty between Ethiopia and the United Kingdom, relative to the frontiers between the Anglo-Egyptian Sudan, Ethiopia, and Eretria', concluded in 1902 in Addis Ababa (Great Britain–Ethiopia Treaty 1902). The Treaty stated in Article III that Emperor Menelik II of Ethiopia:

> engages himself towards the Government of His Britannic Majesty not to construct, or allow to be constructed, any work across the Blue Nile, Lake Tsana, or the Sobat which would arrest the flow of their waters into the Nile, except in agreement with His Britannic Majesty's Government and the Government of the Sudan.
>
> (Great Britain–Ethiopia Treaty 1902: Article III)

The treaty has been, for a long time, a source of a bitter dispute between Ethio-pia and Egypt over the Nile. Ethiopia vehemently rejects this treaty, claiming that the treaty was not ratified by any of the government organs, and that the Amharic and English versions of the treaty are different with respect to the said article (Degefu 2003). Ethiopia even contends that Egypt is not a party to this treaty, and even if the treaty is valid, Egypt cannot make any claims therefrom (Yihdego 2017). On the other hand, Egypt insists that the treaty is valid and binding on Ethiopia, and that according to its provisions, Ethiopia cannot build any project on the Nile prior to Egypt's agreement, as a successor to the treaty. Thus, this treaty is a major source of dispute between Ethiopia on the one hand, and Egypt and Sudan on the other.

Another treaty that is a major source of dispute is the 1929 Nile Waters Agreement concluded by Britain on behalf of its East African colonies of Kenya, Uganda and Tanganyika, as well as the Sudan, on the one hand, and Egypt on the other (Great Britain Egypt Agreement 1929). Paragraph 4 (ii) of the Agreement stated that:

> Except with the prior consent of the Egyptian Government, no irrigation works shall be undertaken nor electric generators installed along the Nile and its branches nor on the lakes from which they flow if these lakes are situated in Sudan or in countries under British administration which could jeopardize the interests of Egypt either by reducing the quantity of water flowing into Egypt or appreciably changing the date of its flow or causing its level to drop.
>
> (Great Britain Egypt Agreement 1929: Paragraph 4 [ii])

Upon attaining independence in the early 1960s, Kenya, Uganda and Tanganyika (which was succeeded by Tanzania, as a result of the unification of Tanganyika and Zanzibar in 1964) argued that they are not bound by this agreement because they were not parties to it (Garretson 1967). These countries also invoked the Nyerere Doctrine (named after Julius Nyerere, the first prime minister, and later president of Tanganyika, and thereafter Tanzania), under which Mr Nyerere gave the treaties concluded during the colonial era two years to be renegotiated, after which they would lapse, if no new treaty was reached (Mekonnen 1984).

Egypt, on the other hand, invoked the principle of state succession to support its claim that the 1929 agreement remains valid and binding. Egypt claims that, like the 1902 treaty with Ethiopia, the agreement gives Egypt veto power over any project on the Nile that would jeopardise the interests of Egypt (Tvedt 2004).

A third treaty that is a source of a bitter dispute between Egypt and Sudan on the one hand, and the other Nile riparians on the other hand, is the 1959 Nile Waters Agreement (Egypt–Sudan Agreement 1959). The agreement was concluded in Cairo on 8 November 1959 and entered into force on 22 November 1959. The agreement established the total annual flow of the Nile (measured at Aswan) as 84 billion cubic metres (BCM), and allocated 55.5 BCM to Egypt and 18.5 BCM to the Sudan. The remaining 10 BCM represent the annual evaporation and seepage losses at the large reservoir created by, and extending below, the Aswan High Dam in southern Egypt and northern Sudan. Thus, Egypt and Sudan allocated the entire flow of the Nile, as measured at Aswan, to themselves. Indeed, the agreement itself is titled, 'Agreement between the

United Arab Republic and the Republic of the Sudan for the Full Utilization of the Nile Waters'.

The agreement sanctioned the construction of the Aswan High Dam in Egypt, and the Roseires Dam on the Blue Nile in Sudan. The Agreement established the Permanent Joint Technical Committee, with an equal number of members from each country, as the bilateral institutional mechanism for the joint management of the Nile.

While Egypt and Sudan recognised the claims of the other riparian states to a share of the Nile waters, if the other states so requested, they reserved to themselves the ultimate decision on whether these states would get a share, and if so, how much. They further entrusted the Permanent Joint Technical Committee with supervision of the use of any amount so granted to any such riparian by Egypt and Sudan. In support of their water allocation under the 1959 Agreement, Egypt and Sudan claim historic and established rights to the waters of the Nile (Krishna 1988).

The agreement also stipulated that in case any question connected with the Nile waters needs negotiations with the governments of any riparian territories outside the Republic of Sudan and Egypt, the two republics shall agree beforehand on a unified position in accordance with the investigations of the problem by the Permanent Joint Technical Committee. This unified position shall form the basis of instructions to be followed by the Committee in the negotiations with the governments concerned.

Ethiopia had demanded participation in the 1959 negotiations, but Egypt and Sudan ignored that request. Consequently, Ethiopia sent a number of memoranda rejecting the agreement, and Britain, on behalf of Kenya, Uganda and Tanzania followed the same route. Egypt and Sudan now claim that their share of the Nile waters under the 1959 Agreement has become an established and acquired right, and the other riparians are obliged under international law not to interfere with it, as this would cause significant harm to Egypt and Sudan.

The 1959 Nile Waters Agreement is totally rejected by the other riparian states, who argue that they are not parties to the agreement, and have never acquiesced to it. As riparians, they consider the claims of Egypt and Sudan to the entire flow of the Nile as an infringement of their rights under international law for a reasonable and equitable share to the Nile waters, given that the entire flow of the Nile originates within their territories (Okidi 1994). These riparians also raise the principle of equality of all the riparians of the shared watercourse as the basic principle of international water law. For Egypt and Sudan, the 1959 Agreement basically sealed their rights on the Nile waters, and confirmed the veto power under the 1902 and 1929 agreements (Swain 1997).

However, these agreements, and Egypt's and Sudan's contentions, have been heavily shaken through time and projects in the upstream riparian countries, and through subsequent agreements on the Nile waters, as discussed below.

Post-colonial agreements: the 1991 and 1993 Nile Waters Agreements

The collapse of the Marxist regime of Mengistu Haile Mariam in Ethiopia in May 1991 brought Ethiopia to the forefront of the search for solutions to the Nile disputes, after years of having participated as an observer in the Nile meetings.[1] The new leader, Mr Meles Zenawi, who came to power in May 1991, visited Khartoum in December 1991, and concluded a series of agreements on various matters with the Sudanese government on 23 December 1991, under the umbrella 'Ethiopia – Sudan Peace and Friendship – Khartoum Declaration' (Ethiopia Sudan Declaration 1991).[2]

Paragraphs 4.1.1 to 4.1.7 of the Declaration dealt with 'Minutes of Appreciation of Principles and Needs for Uses of the Nile Waters'. The first paragraph recognised that: 'the Nile is a common resource of the co-basin states; a vital resource, one of which our lives, and respective cultures are intimately tied to it forever'. The paragraph added that: 'the two sides believe in, and affirm equitable entitlements to the uses of the Nile Waters without causing appreciable harm to one another'. Thus, Paragraph 1 of the Declaration restated the basic principle of international water law of the equality of all the riparians, and also referred to the two other principles of equitable and reasonable utilisation, and the obligation not to cause harm. The Declaration then went on to address the issues of soil erosion, flooding and sediments influxes, and the need for natural resources inventory.

In Paragraph 5, the two parties acknowledged the absence of Ethiopia from the previous initiatives on the Nile, and agreed that: 'Ethiopia will now declare upgraded participation at a full membership level to contribute for the regional and international initiatives that tend to forge cooperation among the basin countries.'

Paragraph 6 of the Declaration stated that: 'it is deemed essential to establish a joint technical committee with multi-discipline composition of experts who would pave the way for prior consultations; exchange of data and information; and explore areas of cooperation'. This committee, as stated in the last paragraph of the Declaration, would be a first step 'towards the objective of achieving the formation of the Nile Basin Organization, taking the interests of all riparian countries, and with their universal consent'.

Thus, Sudan accepted the basic principles of international water law of equality of all the riparians, cooperation, equitable and reasonable utilisation, and the obligation not to cause harm, as well as the need for a basin-wide organisation for the Nile.

Less than two years later, Ethiopia concluded on 1 July 1993, a similar agreement with Egypt, titled 'Framework for Regional Cooperation between Ethiopia and the Arab Republic of Egypt' (Ethiopia Egypt Framework Agreement 1993).[3] The first two paragraphs of the Framework Agreement dealt with the principles of good neighbourliness, peaceful settlement of disputes and non-interference in the internal affairs of states, as well as the consolidation of mutual trust and understanding between the two countries. Article 3 stated the recognition of the importance of cooperation of the two parties as an essential means to promote their economic and political interests, as well as stability of the region.

The remaining five articles of the Framework Agreement (4 to 8) addressed specifically the Nile Basin. Article 4 of the Framework Agreement stated that: 'the issue of the use of the Nile waters shall be worked out in detail through discussions by experts from both sides, on the basis of the rules and principles of International Law'.

The two parties agreed under Article 5 to refrain from engaging in any activity related to the Nile waters that may cause appreciable harm to the interests of the other party. This article reflects the different respective thinking by Egypt and Ethiopia about harm. While Egypt seemed to think along the classic, and erroneous, line that harm could only be caused by upstream riparians to the downstream ones, Ethiopia has taken the correct view that harm is indeed a two-way matter; and just as upstream riparians can harm downstream riparians, downstream riparians can also cause harm to upstream riparians.

It is obvious, and clearer, that the downstream riparians can be harmed by the physical impacts of water quantity and quality changes caused by water use by the upstream riparians. The quantity of water flow can be decreased by the upstream riparians through construction of dams, canals and pipelines, and through the storage and diversion of the waters of the shared rivers. The quality of the water of the shared rivers can be affected by the upstream riparians through pollution caused by industrial waste, sewage or agricultural run-off.

It is much less obvious, and generally not realised, that the upstream riparians can be affected, or even harmed, by the potential foreclosure of their future use of water, caused by the prior use, and the claiming of rights to such water by the downstream riparians. Projects on shared rivers in the downstream riparian states would help those riparians in acquiring, and later claiming, rights to the water abstracted under those projects. The availability and use of such waters in future by the upper riparians would have already been foreclosed by the downstream riparians. Such downstream riparians would usually invoke the principle of acquired rights and the obligation not to cause harm in the face of claims by the upstream riparians. The upstream riparians' claims to the part of the waters of the shared watercourse under the principle of equitable and reasonable utilisation would be countered by claims of historic rights and existing uses, and the obligation against causing harm by affecting them (Salman 2010). Thus, as Ethiopia is under obligation not to cause significant harm to Egypt and the Sudan, Egypt and the Sudan are equally obliged not to cause significant harm to Ethiopia through foreclosure of Ethiopia's future uses.

Indeed, Ethiopia indicated its understanding of the concept of foreclosure of future uses in its Note Verbale of 20 March 1997, addressed to Egypt, on the Toshka or New Valley Project which Egypt was constructing, and which draws water from the Nile River. The Note Verbale stated:

> Ethiopia wishes to be on record as having made it unambiguously clear that it will not allow its share to the Nile waters to be affected by a *fait accompli* such as the Toshka project, regarding which it was neither consulted nor alerted.
>
> (Waterbury 2002)

Thus, it is clear that each of Egypt and Ethiopia has it views on the obligation against causing of significant harm, although the Ethiopian understanding is legally the correct one.

Article 6 of the Framework Agreement dealt with the necessity of conservation and protection of the Nile waters, and the need to consult and cooperate: 'in projects that are mutually advantageous, such as projects that would enhance the volume of flow and reduce the loss of Nile Waters through comprehensive and integrated development schemes'.

Article 7 expressed the desire of the two parties to create an appropriate mechanism for periodic consultations on matters of mutual concern, including the Nile waters, in a manner that would enable them to work together for peace and stability in the region. In Article 8, the two parties agreed to: 'endeavor towards a framework for effective cooperation among the countries of the Nile Basin for the promotion of the common interest in the development of the basin'.

Thus, the Ethiopian Egyptian Framework Agreement dealt with good neighbourliness and peaceful settlement of disputes. It also addressed the basic issues of cooperation and the need to establish appropriate mechanism for consultation, as well as the obligation not to cause harm, as discussed above.

It is noteworthy that neither the Declaration, nor the Framework Agreement, included any solid explicit plans for cooperative projects and programmes. Yet both instruments are important indicators in recognising Ethiopia's rights in the Nile waters by Egypt and Sudan. This is a significant reversal of the position of the two countries under the 1959 Nile Waters Agreement where Egypt and Sudan divided the entire Nile flow between them, leaving no share for the other Nile riparians. Also, as discussed above, under the same Agreement, Ethiopia, the source of 86 per cent of the Nile waters is required under the 1959 Nile Waters Agreement to apply to Egypt and Sudan for a share in the Nile waters, and the two countries gave themselves the right to accept or reject that application. If they accept the application, they would decide how much Nile waters should be allocated to Ethiopia, and have mandated the Permanent Joint Technical Committee to ensure that the allotted amount is not exceeded.

More importantly, however, is the fact that neither the Declaration, nor the Framework Agreement, made any explicit or implied reference to the 1902 Treaty, or to the claim of Egypt of veto power over Ethiopia's projects and programmes on the Nile.

One significant observation under the Declaration was the agreement of Ethiopia to upgrade its participation to a full membership level to contribute in the regional and international initiatives that tend to forge cooperation among the basin countries. Until that time, Ethiopia tended to participate as an observer, or not to participate at all, in those meetings and initiatives (NBI 2016). The upgrade was a major building block in the establishment of the Nile Basin Initiative (NBI), which emerged in the late 1990s, as discussed below.

Genesis of the attempts towards cooperation

Despite the disputes over the 1902 Treaty, the 1929 Agreement, and the 1959 Agreement, the Nile riparians started looking at ways for working together. This process began with the meetings in the mid-1960s to discuss the rising levels of Lake Victoria. The process led to the Hydromet project, which was facilitated and funded by some United Nations agencies. There were speculations that the rising levels of Lake Victoria might have been caused by the Sudd swamps of South Sudan, or/and by the Aswan High Dam that was about to be completed at that time. Accordingly, Egypt and Sudan were invited, and they happily joined the Hydromet project.

The process continued and proceeded through the Undugu Group (Swahili word for brotherhood) in the 1970s and 1980s; TECCONILE; and the Nile 2002 conferences in the 1990s (Brunnée and Toope 2002). However, these efforts did not go beyond attempts to improve communication between the Nile riparians, at the experts' level.

Building on these efforts, the World Bank and the United Nations Development Programme (UNDP), together with some other donors, started in 1997 to facilitate the establishment of a more formal setting for cooperation among the Nile Basin riparians. This mechanism was called the Nile Basin Initiative (NBI). After a series of informal meetings, the NBI was officially established by the Nile Basin states at the meeting of their Ministers of Water Resources held in Dar es Salaam, Tanzania, on 22 February 1999. The Agreed Minutes of the meeting, signed by the ministers in attendance, included the overall framework for the NBI and its institutional structure and functions. All the Nile riparian states at that time became members of the NBI, except Eritrea who opted for being an observer.

The NBI is established as an intergovernmental organisation, and has been viewed as a transitional arrangement to foster cooperation and sustainable development of the Nile River for the benefit of the inhabitants of those countries. It was able to bring together, for the first time, the ten riparian states (at that time) at the ministerial level. The NBI is guided by a shared vision 'to achieve sustainable socio-economic development through equitable utilization of, and benefit from, the common Nile Basin water resources' (NBI 2012b). As will be discussed later, the emphasis on equitable utilisation is quite noteworthy.

The NBI is managed by a transitional institutional structure, including the NBI Secretariat located in Entebbe, Uganda, and two project offices, one in Addis Ababa, Ethiopia for the Eastern Nile, and the other in Kigali, Rwanda, for the Southern Nile (also called Nile Equatorial Lakes). In addition, the organs of the NBI include the Council of Ministers of Water Resources of the

Nile Basin Countries (Nile-COM), which provides policy guidance and makes decisions on major matters relating to the NBI (NBI 2016).

Consequent to the establishment of the NBI, the Nile Basin Initiative Act was enacted by the Republic of Uganda in October 2002 to give the force of law in Uganda to the signed Agreed Minutes of the meeting of the Nile-COM, held in Cairo, Egypt, on 14 February 2002. According to the Agreed Minutes, the ministers decided to: 'invest the NBI, on a transitional basis, with legal personality to perform all of the functions entrusted to it' (Salman 2009). The Headquarters Agreement was concluded between the Government of the Republic of Uganda and the NBI in Kampala on 4 November 2002, following the enactment by Uganda of the Nile Basin Initiative Act.

The Nile Basin Trust Fund was thereafter established for financing joint projects in the Nile riparian states, with funding from a number of multilateral and bilateral donors. The World Bank, which has spearheaded the mediation efforts of the NBI, has also administered the Trust Fund.

With the transitional institutional structure for the NBI, and the Trust Fund, in place, the Nile Basin countries directed their attention to the main task they undertook to perform – preparation of an inclusive legal instrument for the Nile Basin.

The Nile Basin Cooperative Framework Agreement

As discussed above, the main objective of the NBI has been to conclude a cooperative framework agreement that would incorporate the principles, structures and institutions of the NBI, and that would be inclusive of all the Nile riparians. Work on the Nile Basin Cooperative Framework Agreement (CFA) started immediately after the NBI was formally established in 1999, and continued for more than ten years. However, the process ran into some major difficulties as a result of the resurfacing and hardening of the respective positions of the Nile riparians over the colonial treaties, as well as the Egyptian and Sudanese claims to what they see as their acquired uses and rights of the Nile waters.[4] Those differences, as discussed later, persisted and could not be resolved at the negotiations level and were eventually taken to the ministerial meetings that took place in Kinshasa, Alexandria and Sharm el-Sheikh in 2009 and 2010. However, those meetings in turn failed to resolve those differences and no agreement on the final draft CFA could be reached.

On 14 May 2010, four of the Nile riparians (Ethiopia, Tanzania, Uganda and Rwanda) signed the CFA in Entebbe, Uganda, and were joined five days later by Kenya (CFA 2010). On 28 February 2011, Burundi joined those five states and signed the CFA. Although the Democratic Republic of Congo indicated its support of the CFA, it has not yet signed the CFA. The CFA is also referred to as the 'Entebbe Treaty' or 'Entebbe Agreement' because it was signed at Entebbe, Uganda.

The CFA lays down some basic principles for the protection, use, conservation and development of the Nile Basin. The CFA established the principle that each Nile Basin state has the right to use, within its territory, the waters of the Nile River Basin, and lays down a number of factors for determining equitable and reasonable utilisation. In addition to the factors enumerated in the United Nations Watercourses Convention,[5] the CFA factors included the contribution of each Basin state to the waters of the Nile River System, and the extent and proportion of the drainage area in the territory of each Basin state.

The CFA also included provisions requiring the Nile Basin states to take all appropriate measures to prevent the causing of significant harm to other Basin states. The CFA indicated further that where significant harm nevertheless is caused to another Nile Basin state, the states whose use causes such harm shall, in the absence of agreement to such use, take all appropriate measures, having due regard for the provisions of the CFA on equitable and reasonable utilisation, in consultation with the affected state, to eliminate or mitigate such harm, and, where appropriate, to discuss the question of compensation. These provisions are exactly the same as those of the UN Watercourses Convention.

The articles of the CFA on equitable and reasonable utilisation and the obligation against causing significant harm resulted in the same discussion and controversies that the respective provisions of the Watercourses Convention generated in the 1990s. It is noteworthy that, as a general rule, the lower riparians favour the no-harm rule, since it protects their existing uses against impacts resulting from activities undertaken by upstream states. Conversely, upper riparians favour the equitable utilisation principle because it provides more scope for states to utilise their share of the watercourse for activities that may impact on downstream states (Salman 2007b).

However, it is widely believed that the Watercourses Convention has resolved this controversy by subordinating the obligation against causing significant harm to the principle of equitable and reasonable utilisation (Salman 2007a). This belief is supported and strengthened further by the decision of the International Court of Justice in the Danube case where the Court emphasised repeatedly the principle of equitable and reasonable utilisation, and made no reference to the obligation against causing significant harm (Gabcikovo Nagymaros 1997). Indeed, the differences between Egypt and Sudan on the one hand, and the upstream riparians on the other, on the CFA, stem from, and are centred around this area, and are interwoven with the issues related to the 1902 and 1929 agreements, as well as the claims of Egypt and Sudan under the 1959 Nile Agreement, as discussed below.

Areas of differences over the Cooperative Framework Agreement

In an attempt to address the controversies and differences over these two principles between Egypt and Sudan on the one hand, and the remaining riparians on the other, the CFA facilitators and negotiating teams introduced the concept of water security. Article 2 of the CFA defined water security to mean: 'the right of all Nile Basin States to reliable access to and use of the Nile River system for health, agriculture, livelihoods, production and the environment'. Article 14 required the Basin states to work together to ensure that all states achieve and sustain water security. However, these paragraphs did not satisfy Egypt and Sudan who want to ensure, through an additional clause, that their existing uses and rights are fully protected under the CFA.

Egypt and Sudan demanded and insisted that Article 14 of the CFA should include a specific provision, to be added at the end of the Article, which would oblige the Basin states: 'not to adversely affect the water security and current uses and rights of any other Nile Basin State'. The position of Egypt and Sudan revived the long-standing disputes related to the treaties discussed above, namely Egypt's veto power, and Egyptian and Sudanese claims to their existing uses of, and rights to, the Nile waters under the 1959 Nile Agreement.

It should be added that Egypt and Sudan signed off on the objective of the NBI, which is 'to achieve sustainable socio-economic development through equitable utilization of, and benefit from, the common Nile Basin water resources'. Nevertheless, the two countries insisted throughout the subsequent entire negotiations process of the CFA on their existing uses and rights, and this is basically a claim to the no-harm rule. This position is tantamount to a demand for an unequivocal recognition of the 1902 and 1929 agreements, and Egypt's veto power therein. However, the upper riparians do not see the demands of Egypt and the Sudan as consistent with the vision of the NBI, or with international law. No wonder that the demand of Egypt and Sudan was totally rejected by the upper riparian states.

Thus, rather than assist in the resolution of the controversies over the principle of equitable and reasonable utilisation and the obligation against causing significant harm, introduction of the third concept of water security simply widened the gap and exacerbated the differences over the two principles, between the Nile lower and upper riparians, and over on the CFA as a whole (Mekonnen 2010).

Indeed, the concept of water security turned out to be an unnecessary and unfortunate addition to the CFA, and actually a complicating factor. It is more or less a redundant reiteration of the principle of equitable and reasonable utilisation, and the obligation against causing significant harm. The Nile upper riparians went ahead with it because it reinforced their belief that the CFA subordinates the no-harm rule to the equitable and reasonable utilisation principle. On the other hand, Egypt and Sudan thought they would use it to inject the new clause to protect fully their existing uses and rights.

As a result, the new concept of water security simply reopened the major pre-NBI differences on colonial treaties and existing uses and rights, and derailed the CFA negotiations process. It is worth adding that this concept does not exist in any other multilateral or bilateral water treaty, and is not a legal theory per se, but is merely a political concept (Salman 2013).

Besides the issue of water security, Egypt and Sudan demanded that the CFA should include explicit provisions on notification of other riparians of planned measures, which may cause significant adverse effects to other riparians. The

CFA required the Nile Basin states to exchange information on planned measures through the Nile River Basin Commission to be established under the CFA, and included no provisions on notification.

Ethiopia opposed notification of other riparians because of its concerns that such notification may be construed as recognition of the 1902 Treaty, and could give Egypt the opportunity to reassert its claim of veto power over Ethiopia's projects on the Nile, as discussed earlier. Ethiopia further alleges that it was not notified by Egypt and Sudan of any project affecting the Nile, and so it is not obliged, under the reciprocity rule, to notify them (Salman 2010).

In support of their claim, Egypt and Sudan demand that provisions on notification of other riparians, in line with those included in the World Bank Policy for Projects on International Waterways,[6] and the UN Watercourses Convention, be included in the CFA.

It is worth noting in this regard that neither Egypt, nor Sudan, nor any of the Nile riparian states is a party to the Watercourses Convention. The Convention was adopted by the United Nations General Assembly on 21 May 1997, by a vote exceeding 100 members. It entered into force and effect on 17 August 2014, sixty days after Vietnam deposited the 35th instrument of ratification/accession on 19 May 2014 (Salman 2015). Kenya and Sudan voted for the Convention, while Burundi voted against it. Egypt, Ethiopia, Rwanda and Tanzania abstained, each for its own reasons, while Eritrea, the Democratic Republic of Congo and Uganda did not participate in the vote. Thus, the invocation by Egypt and Sudan of the Convention's provisions is quite interesting, given that neither of them is a party to the Convention.

Another major difference between the parties relates to the manner in which the CFA would be amended. While Sudan and Egypt demand that the CFA should be amendable by consensus, or a majority that includes both Egypt and Sudan, the other riparians insist that the CFA be amendable by a simple majority of the riparian states, whether it includes Egypt or Sudan or not, with no veto power to any riparian.

It should be added that the CFA distinguishes between the 'Nile River Basin' and 'Nile River System'. It defines the 'Nile River Basin' to mean: 'the geographical area determined by the watershed limits of the Nile River System of waters; this term is used where there is a reference to environmental protection, conservation or development'. The CFA defines the 'Nile River System' to mean: 'the Nile River and the surface waters and groundwaters which are related to the Nile River; this term is used where there is a reference to utilization of water'. No conclusive agreement has been reached on those definitions. Egypt has been claiming that the definition of the term 'Nile River System' applies to both: environmental protection and water allocation. This would mean, according to Egypt, that the waters of the Nile would include not only the 84 BCM measured at Aswan and specified by Egypt and Sudan in the 1959 Nile Agreement, which the two countries currently fully utilise (blue water), but also the more than 1,600 BCM of rainwater that Egypt claims falls annually on the basin area (green water). In other words, Egypt would want to include not only the waters that flow in the river (blue water), but the rain that falls on the entire basin area (green water).[7] This would be an increase of close to 1,800 per cent of the Nile current flow of 84 BCM measured at Aswan. Under this scenario, Egypt's current uses would be a small fraction of the 'Nile River System'. If Egypt insists on this definition, it would certainly infuriate the other riparians who consider this to be an absurd definition of the Nile River System. It would add another major complication to the CFA negotiations, and to the already difficult relations between the Nile upper and lower riparians.

Thus, a wide array of major issues divide the two groups of Nile riparians. This intricate situation was exacerbated further by the February 2011 signature by Burundi as the sixth riparian of the CFA, as mentioned earlier, rendering the entry into force of the CFA a possibility. This is because the CFA needs, according to its own provisions, six ratifications to enter into force and effect.

The emergence of the Republic of South Sudan as an independent state on 9 July 2011, and its admission to the NBI on 5 July 2012, is another development that may favour the upper riparians. South Sudan is expected, based on ethnicity, geography, history, culture and interests, to side with the Equatorial Lakes' countries and Ethiopia, and to adopt their stance on the Nile Basin and the CFA issues (Salman 2011b).

However, entry into force of the CFA with only the six countries that have thus far signed it, or even seven riparian states, would result in some major legal problems and complications. The CFA establishes the Nile Basin Commission for promoting and facilitating the implementation of the principles, rights and obligations set forth in the CFA, and for serving as an institutional framework for cooperation among the Nile Basin states. Article 30 of the CFA states that upon entry into force of the CFA, the Commission shall succeed to all rights, obligations and assets of the NBI. The immediate question that arises is what will happen to the rights and obligations under the NBI of the states that are not parties (and do not plan to be parties) to the CFA? This situation, if it does materialise, will raise some difficult legal issues, and will exacerbate the existing disputes over the CFA. Clearly, the mediators and drafters of the CFA neither anticipated, nor took into account, the occurrence of this situation.

Moreover, the challenges and difficulties over the sharing of the Nile waters, as reflected in the differences over the CFA, would, a year after the signature of the CFA in 2010, be exacerbated by another major development, the Grand Ethiopian Renaissance Dam (GERD), as discussed below.

Challenges and opportunities of the Grand Ethiopian Renaissance Dam

As if the differences over the CFA were not enough, another major dispute erupted in March 2011, about ten months after the signature of the CFA in May 2010. Ethiopia announced at the end of March 2011 that it has planned to build the Grand Ethiopian Renaissance Dam (the GERD, known formerly as the Grand Millennium Dam), and indeed, construction started at the beginning of April 2011. The GERD is being built on the Blue Nile, about 20 kilometres from the Sudanese borders, with a generating capacity of 6,000 megawatts (raised recently to 6,450), a reservoir size of 74 BCM, and at a cost close to $5 billion. Ethiopia claims that the dam will benefit Egypt and Sudan through flood and sediments control, and regulation of the river flow, and will generate electricity at a low cost that could be sold to the other Nile riparians, including Egypt and Sudan.

Immediately after the announcement was made, Egypt and Sudan vehemently opposed the dam, claiming adverse effects to their Nile water rights and interests. They asked Ethiopia to provide them with all the studies on the dam, as part of the notification process, so that they could determine the effects of the GERD on both of them.

It should be added that the GERD is not the first dam to be built by Ethiopia on the Nile. Ethiopia started with the Fincha Dam – a small dam on the Blue Nile that was completed in 1973 for generation of hydropower. It was followed by two small dams, the first and second Tis Abbay projects, on Lake Tana for generation of hydropower also. However, at the beginning of this century, Ethiopia moved to the fourth project on the Nile, the largest and most significant of all. That was the Tekeze Dam on the Atbara River, which was completed in 2010. This large dam, with a height of 188 metres and a storage capacity of 10 BCM, generates more than 360 megawatts of electricity. Concurrently, Ethiopia also built the Tana–Beles project that connected the Beles River (a tributary of the Blue Nile) with Lake Tana, and generated hydropower at that connecting point. The project was also completed in 2010. The last two dams generated the incentives and appetite for Ethiopia to push ahead with the construction of the GERD. These two projects also marked the Ethiopian successful practical challenge to the 1902 Agreement referred to earlier. Ethiopia is also planning on building a number of dams, including Karadobi, Mandaya, Didessa and Mabil on the Blue Nile (World Bank 2006).

A series of high-level meetings took place on the GERD between Ethiopia, Egypt and the Sudan. Consequently, the three countries agreed, on 5 December 2011, to establish the International Panel of Experts to assess the impact of the GERD on the Blue Nile, and on Sudan and Egypt. The Panel consisted of ten members, two experts from each of the three countries, plus four other international experts (International Panel of Experts 2013). This was viewed as a positive step, indicating a major compromise on both parts on the issue of notification, and an attempt to resolve the dispute over the GERD in a peaceful manner.

Sudan, however, later softened its position, and announced in December 2013 that it supports the GERD. This position, together with the series of high-level meetings between the three parties, led to the conclusion of the Agreement on Declaration of Principles on the GERD (DoP) (DoP 2015).

The DoP was no doubt a major breakthrough, not only on the GERD, but also on the entire relationship of the three countries over the Nile. Egypt and Sudan recognised explicitly, and for the first time, the rights of Ethiopia on the

Nile waters. This recognition is included in the preamble of the DoP where the rising demands of the three countries on their transboundary water resources, and the significance of the Nile as the source of livelihoods and development for the people of the three countries, are explicitly acknowledged.

Article 1 of the DoP stipulated the intention of the three parties to cooperate based on common understanding, mutual benefit, good faith, win–win, and the principles of international law, as well as to cooperate in understanding upstream and downstream needs in its various aspects.

Article II stipulated the explicit acceptance of the GERD by Egypt and Sudan. The article stated that the purpose of the GERD is power generation, contribution to economic development and regional integration.

Articles III and IV dealt with the obligation not to cause significant harm, and the principle of equitable and reasonable utilisation. Article V underscored the agreement of the parties to cooperate in the first filling and operation of the GERD, whereas under Article VI priority is to be given to the downstream countries in the purchase of the power generated by the GERD. Article VII dealt with the principle of exchange of information and data, while Article VIII dealt with the principle of dam safety.

Article IX reverted to cooperation between the three countries on the basis of sovereign equality, territorial integrity, mutual benefit and good faith, while Article X dealt with the principle of peaceful settlement of disputes (Salman 2016).

It should be noted that the DoP was signed in Khartoum by the three top political figures in the three countries – Presidents Abdel Fattah El-Sisi of Egypt, and Omar Hassan Ahmad al-Bashir of Sudan, and Prime Minister Hailemariam Desalegn of Ethiopia.

The principle of quality of all the riparians, as pronounced by the International Court of Justice in the Gabcikovo–Nagymaros case, has clearly been accepted and underscored by Egypt and Sudan for the first time ever in the history of the Nile. This is a major and significant reversal of the stipulations concluded by Egypt and Sudan under the 1959 Nile Waters Agreement. Noteworthy also is the absence in the DoP of any reference to the 1902 Treaty, or the 1959 Agreement, or to the demands of Egypt and Sudan regarding recognition of their existing uses and rights.

Indeed, acceptance of these principles by Egypt and the Sudan have been confirmed by another instrument. On 28 December 2015, the three ministers of Water Resources and those of Foreign Affairs signed a document titled 'Summary and the Outcomes of the Meeting' (referred to as the 'Khartoum Document'). The second paragraph of the Preamble reconfirmed 'the sincere and full commitment of the three countries to adhere to the Agreement on Declaration of Principles (DoP)', putting an end to the rumours about Egypt's imminent move to withdraw from the DoP. Thus, the DoP received a major boost, and a clear and unequivocal reconfirmation of its provisions, through the Khartoum Document that was signed nine months after the DoP was concluded and signed by the three top political figures in the three countries. Clearly, and without any doubt, the contours of hydropolitics and law of the Nile are rapidly changing.

Can these developments assist in resolving the differences over the CFA? This is what we will be discussed in the next part of this chapter.

The Nile mediation efforts and the Basin's new realities

The above discussion indicates that the results of the mediation efforts on the Nile spearheaded by the World Bank and UNDP are still uncertain, and far from successful. After more than seventeen years of mediation, an agreement on the CFA has been difficult to reach and the issues are still unresolved. The basic and original differences on the colonial treaties, and acquired uses and rights vis-à-vis equitable and reasonable utilisation still persist. They are compounded by introduction of the unfortunate concept of water security, as well as differences on notification, and on consensus *versus* majority on the amendment of the CFA. Another complication factor was Egypt's insistence on defining the Nile waters as including also the rainwater that falls on the entire basin area, which Egypt claims exceeds 1,600 BCM.

Moreover, the latter part of the negotiation process on the CFA showed an emerging trend of the unified and challenging stance of the upper Nile riparians vis-à-vis Egypt and the Sudan. During the eighteenth annual meeting of the Nile Council of Ministers (Nile-COM) held in Addis Ababa in July 2010, Egypt and Sudan demanded that an extraordinary meeting of the Nile-COM be convened to discuss the CFA, but the upper riparians showed no interest. A similar demand was made a year later in Nairobi during the nineteenth annual meeting of the Nile-COM. However, it was agreed during the Nairobi meeting, as a compromise, that the extraordinary meeting would be held in Kigali on 28 October 2011, but would limit itself to discussing the legal and institutional ramifications of the entry into force of the CFA.

Despite their request, Egypt and Sudan later asked for postponement of the Kigali meeting, ostensibly because of the Egyptian elections. It was consequently agreed that the meeting would take place in Nairobi, Kenya, on 18 December 2011. Yet, again Egypt and Sudan asked for a postponement, and the other riparians reluctantly agreed to this request, rescheduling the meeting to take place on 27 January 2012, in Nairobi, Kenya.

The Nile ministers from the upstream countries did arrive in Nairobi for that meeting, but the Egyptian and Sudanese ministers failed to show up. Consequently, the Nile-COM meeting did not take place. However, the ministers from the equatorial lakes countries convened their own Nile Equatorial Lakes Council of Ministers (NEL-COM) meeting, and agreed to upgrade the observer status of Ethiopia in the NEL-COM to full membership. As a result of this decision, the six countries that signed the CFA succeeded in establishing an institutional framework for working together and for taking collective decisions. The outcome of the NEL-COM meeting was included in the Nairobi Statement, which the ministers issued at the end of the meeting (Salman 2012b).

Annoyed by what they saw as dilatory tactics by Egypt and Sudan, the six ministers of the countries that signed the CFA stated that they would proceed with the process of ratification of the CFA, and the establishment of the Nile Basin Commission once the CFA enters into force. The ministers instructed the Chair of the Nile-COM to continue discussions with the three Nile states that have not signed the CFA (Egypt, Sudan and the Democratic Republic of Congo) with the view of getting them to join the CFA, and decided that such discussions should be concluded within sixty days. The Nairobi Statement indicated the frustrations of the ministers at Egypt and Sudan's indecisiveness regarding the extraordinary meetings that resulted in a number of meeting cancellations on short notice. The Nile-COM chair was instructed to communicate these frustrations to Egypt and Sudan, and to convey the outcome of the Nairobi meeting to the respective governments so that the ratification process of the CFA would commence.

That meeting is no doubt a major watershed and a landmark in the history of the Nile Basin. The NEL ministers reversed the previous decision to delay the ratification process of the CFA to give Egypt sometime after the January 2011 revolution. The discussions to be conducted by the Nile-COM chair with Egypt and Sudan would not be about the areas of disagreements on the CFA, not even about the legal and institutional ramifications of the entry into force of the CFA. They would be about bringing these states to join the CFA, and those discussions should be completed within sixty days.

Thus, the upper riparians are clearly emerging as a power to reckon with. They are asserting their demands for an equitable share of the Nile waters, and are openly challenging the legal and political hegemony of Egypt and Sudan over the Nile Basin that had prevailed throughout the last century (Ibrahim 2012). They even started to criticise openly Egypt and Sudan.

It is worth noting in this connection that Ethiopia's main uses of the Nile waters are for the generation of hydropower, as appears from the GERD, with very limited uses for irrigation and water supply. Ethiopia's uses of the Nile waters for hydropower generation are expected to have limited effects on Sudan and Egypt, if the filling of the reservoirs is planned and executed over a reasonable period of time.[8]

Similarly, the uses of Tanzania and Kenya are largely for water supply, such as the Shinyanga project in the former, with minor uses for irrigation in both countries. In addition to minor irrigation and water supply uses, Uganda is building a number of dams for generation of hydropower, but they are all run-of-river, with no storage, and as such have no effects on the other riparians. Thus, it seems that the aim of the upper riparians is to get recognition of their rights to a share of the Nile waters, more than to actually use such waters.

It should also be added in this connection that the Sudd swamps of South Sudan operate as a regulator of the flow of the White Nile. Actually, any uses of the Nile waters upstream of the Sudd would affect mostly the amounts of water spreading across the Sudd, rather than being fully at the expense of the amount of water flowing past the Sudd to Sudan and Egypt (Block and Rajagopalan 1999).

It seems that those basic facts about use of the Nile waters are being gradually realised by Egypt and Sudan, and such facts are increasingly pushing Egypt and

Sudan to think out of the box of the established uses and rights, and veto power.

Moreover, the unified stance of the Nile upper riparians on the CFA, and the support Ethiopia received on the GERD (and its ability to finance it from its own funds), and the absence of support to Egypt on its rejection of the GERD, have been strong indicators for Egypt and the Sudan that the Nile playing field is being levelled. The times of the monopoly over the Nile waters by Egypt and the Sudan are clearly coming to an end.

The initial signs of recognition of these new realities by Egypt and the Sudan appeared in the 1991 and 1993 Framework Agreements between Ethiopia and the Sudan, and Ethiopia and Egypt respectively. The GERD, and subsequently the DoP and Khartoum Document, took this recognition a long way, to the point of parity and equality of the Nile riparian countries.

Conclusion: disentangling the Nile Gordian Knot

The Nile Basin is evidently passing through critical, challenging and changing times. The old order imposed by the lower riparians, particularly Egypt, which monopolised the Nile Basin for centuries, is being gradually and incrementally challenged by new realities and demands from assertive upper riparians. As a result of the half a century old alliance of Egypt and Sudan, established under the 1959 Nile Waters Agreement, a counter-alliance of the upper riparians has been born and reared under the NBI, and institutionalised by the CFA process. Various geopolitical developments have contributed to this situation.

The political turmoil that has engulfed Egypt after the January 2011 revolution has no doubt provided an opportunity for the Nile upper riparians to press with their claims over the Nile waters and to unite behind these claims. Burundi signed the CFA in February 2011, and Ethiopia announced its plans to construct the Renaissance Dam in March 2011; both events took place a few weeks after the eruption of the Egyptian revolution in January 2011 that kept Egypt busy with its internal political turmoil.

The emergence of the upper riparians as a power to reckon with has also been, in my view, an inevitable consequence of a level playing field resulting from the NBI itself. The new alliance of the upper riparians and their unfolding, albeit peaceful, African spring has engendered more balanced power relations vis-à-vis the downstream riparians – Sudan and Egypt. The CFA has ironically, and inevitably, resulted in solidifying the major differences between the upper and lower riparians. The GERD has carried that process a long way, resulting in the recognition by Egypt and Sudan in 2015 of the equality of all Nile riparian countries. This has heavily shaken the 1902, 1929 and 1959 agreements that Egypt and Sudan have been holding on to very strongly, and that they have relied upon to assert their monopoly over the Nile waters. Moreover, the Egyptian–Sudanese alliance itself has been heavily shaken by Sudan's support of the GERD, and its increasing diplomatic and economic ties with Ethiopia.

Yet, the newly emerging power equilibrium within the Nile Basin could as well generate an opportunity for the two Nile groups to compromise, and to realise that there is no alternative to cooperation, and to resolving their differences over the CFA peacefully, now that the differences over the GERD seem to have been, by and large, resolved, and the GERD has become a reality.

Indeed, the opportunity now exists for resolution of the two major differences over the CFA. Egypt and Sudan have accepted, through the DoP, the principle of equality of all the riparians of the Nile. This effectively means that Egypt and Sudan can no longer (and indeed should no longer) insist on their claims for recognition in the CFA of their water security represented in their existing uses and rights. In fact, with the adequate provisions in the CFA on equitable and reasonable utilisation and the obligation not to cause significant harm, the concept of water security is actually quite redundant, and serves no purpose. Egypt and Sudan, can, and should, drop their insistence on recognition in the CFA of their existing uses and rights.

The *quid pro quo* for Egypt and the Sudan would be inclusion in the CFA of provisions on notification, for both downstream as well as upstream riparians, similar to those of the UN Watercourses Convention.[9] The upper riparians need not worry any more that Egypt and Sudan could invoke the 1902 and 1929 agreements, if they were notified by any of them of projects on the Nile. This is because Egypt and Sudan would no longer, under this compromise, be asserting these agreements, as would be evidenced by dropping the relevant paragraph of water security in the CFA. Additionally, all notifications could be handled through the Nile Basin Commission, as was done by the NBI Secretariat (Salman 2009), before the differences within the two groups erupted.

This compromise would effectively address the concerns of both sides to the dispute on the CFA, and would pave the road for genuine and elaborate cooperation by all the riparians on all the Nile issues and challenges, under the wider umbrella of the CFA and its Commission.

Indeed, the areas of cooperation, mutual benefits and inter-dependency in the Nile Basin are quite immense. The potential of hydropower from the Nile in Ethiopia is vast, exceeding 30,000 megawatts. In the Sudan, the current reported irrigated area ranges from 1.2 to 2.2 million hectares, which is about 20 per cent of the total irrigable lands in Sudan. Lake Victoria is rich with a wealth of fisheries, exceeding 500 species, with the current annual catch reaching 500,000 metric tons. The livestock wealth of South Sudan is staggering and has not yet been commercialised, and Egypt food industries are relatively advanced and expandable.

It will only be through genuine cooperation that the above benefits can be harnessed, and the 250 million people who live or depend on the Nile Basin can be pulled from their poverty and underdevelopment. Indeed, with political determination to work in a cooperative and collegial spirit, the Gordian Knot of the CFA can be disentangled for the benefit of the eleven Nile riparian countries and their poor populations who are growing by the day in numbers and poverty.

Notes

1 Ethiopia participated as an observer in the Hydromet project, and the Undugo and Tecconile processes, as will be discussed in the next part of this chapter.

2 The Minutes were signed by Dr Yagoub Abu Shora, Minister of Irrigation and Water Resources of the Sudan, and Mr Aragawa Truneh Kassa, Minister of Housing, Urban Development and Construction of Ethiopia.

3 The Framework Agreement was signed by Mr Meles Zenawi, President of the Transitional Government of Ethiopia, and Mr Hosni Mubarak, President of the Arab Republic of Egypt.

4 Sudan insists on its 'rights' of 18.5 BCM stipulated in the 1959 Nile Agreement because its uses of the Nile waters have only been about 12 BCM annually. On the other hand, Egypt insists on its 'uses' because such uses exceed the 55.5 BCM under the 1959 Agreement. Thus, Sudan and Egypt interests are not actually congruent on this matter (Salman 2014).

5 The factors, enumerated in Article 6 of the Convention and common with those of the UN Watercourses Convention are: (a) geographic, hydrographic, hydrological, climatic, ecological and other factors of a natural character; (b) the social and economic needs of the watercourse states concerned; (c) the population dependent on the watercourse in each watercourse state; (d) the effects of the use or uses of the watercourses in one watercourse state on other watercourse states; (e) existing and potential uses of the watercourse; (f) conservation, protection, development and economy of use of the water resources of the watercourse and the costs of measures taken to that effect; and (g) the availability of alternatives, of comparable value, to a particular planned or existing use.

6 The World Bank has developed, through the years since 1956, elaborate policies for projects on international waterways financed by the Bank, including comprehensive provisions on notification. For a detailed analysis of these policies, see Salman 2009.

7 It is worth mentioning that the Watercourses Convention lists 'the availability of alternatives, of comparable value, to a particular planned or existing use' as one of the factors for determining equitable and reasonable utilisation, and not for allocation of the alternative among the different riparians (McCaffrey 2007).

8 It should be noted that Egypt was able to fill the Aswan High Dam (AHD) reservoir, which has a capacity of 162 BCM, without major negative impact to its huge irrigated agriculture. The capacity of the GERD reservoir is 74 BCM, about 40 per cent of that of the AHD.

9 The Convention requires notification in case a 'project may have significant adverse effect upon other watercourse States,' and lays down detailed provisions on the process. The provisions include furnishing the notified states with project documents, including the environmental impact assessment, and giving them six months to respond. Furthermore, the provisions state that in case of an objection by one or more of the riparian states, the parties should enter into consultations and, if necessary, negotiations with a view to arriving at an equitable resolution of the situation. Thus, the Convention does not grant a veto power to any riparian over planned measures of another riparian (Salman 2015).

References

Block, P. and Rajagopalan, B. (1999). Statistical-Dynamical Approach for Streamflow Modeling at Malakal, Sudan, on the White Nile River. *Journal of Hydrologic Engineering*, 14(2).

Brunnée, J. and Toope, S.J. (2002). The Changing Nile Basin Regime: Does Law Matter?. *Harvard International Law Review*, 43(1), 105–159.

CFA (2010). The Nile Basin Cooperative Framework Agreement. Retrieved from www. nilebasin.org/images/docs/CFA%20-%20English%20%20FrenchVersion.pdf.

Collins, R. (1996). *The Waters of the Nile: Hydropolitics and the Jonglei Canal, 1900–1988*. Princeton, NJ: Markus Wiener Publishers.

Collins, R. (2002). *The Nile*. New Haven, CT: Yale University Press.

Degefu, G.T. (2003). *The Nile: Historical, Legal and Developmental Perspectives*. Victoria, BC: Trafford Publishing.

DoP (2015). Agreement on Declaration of Principles between the Arab Republic of Egypt, the Federal Democratic Republic of Ethiopia, and the Republic of the Sudan on the Grand Ethiopian Renaissance Dam Project (GERDP). Retrieved from http:// hornaffairs.com/en/2015/03/25/egypt-ethiopia-sudan-agreement-on-declaration-of-principles-full-text/.

Egypt–Sudan Agreement (1959). Agreement between the United Arab Republic and the Republic of the Sudan for the Full Utilization of the Nile Waters. *United Nations Treaty Series, 453*(64) (1963).

Ethiopia Egypt Framework Agreement (1993). Retrieved from http://gis.nacse.org/tfdd/ tfdddocs/521ENG.pdf.

Ethiopia Sudan Declaration (1991). Copy is on file with the author.

Gabcikovo–Nagymaros (1997). *Case Concerning the Gabcikovo Nagymaros Project*. Hungary/Slovakia. Retrieved from www.icj-cij.org/docket/files/92/7375.pdf.

Garretson, A.H. (1967). The Nile Basin. In A.H. Garretson, R.D. Hayton and C.J. Olmstead (eds), *The Law of International Drainage Basins*. Dobbs Ferry, NY: Oceana Publications.

Great Britain Egypt Agreement (1929). Exchange of Notes between Great Britain and Northern Ireland and Egypt in Regard to the Use of the Waters of the River Nile for Irrigation Purposes, Cairo, May 1929. *League of Nations Treaty Series* 2103.

Great Britain–Ethiopia Treaty (1902). Treaty between Ethiopia and the United Kingdom, Relative to the Frontiers Between the Anglo-Egyptian Sudan, Ethiopia, and Eretrea, 1902. Retrieved from http://untreaty.un.org/ilc/documentation/english/a_ 5409.pdf.

Ibrahim, A.M. (2012). The Nile Basin Cooperative Framework Agreement: The Beginning of the End of Egyptian Hydro-Political Hegemony. *Missouri Environmental Law and Policy Review, 18*(2), 283–313.

International Panel of Experts (2013). Grand Ethiopian Renaissance Dam, Final Report, Addis Ababa, Ethiopia, 31 May.

Krishna, R. (1988). The Legal Regime of the Nile River Basin. In J. Starr and D. Stoll (eds), *The Politics of Scarcity: Water in the Middle East*. Boulder, CO: Westview Press.

McCaffrey, S. (2007). *The Law of International Watercourses*, 2nd edn. Oxford: Oxford University Press.

Mekonnen, D.Z. (2010). *The Nile Basin Cooperative Framework Agreement Negotiations and Adoption of a 'Water Security' Paradigm: Flight into Obscurity or a Logical Cul-de-Sac?*. European Journal of International Law, 21(2), 421–440.

Mekonnen, Y. (1984). *The Nyerere Doctrine of State Succession and the New States of East Africa*. Arusha, Tanzania: Eastern Africa Publications.

NBI (2012a). *South Sudan Admitted to the Nile Basin Initiative*. Retrieved from www.nile-basin.org/newsite/index.php?option=com_content&view=article&id=127%3Asouth-sudan-admited&catid=40%3Alatest-news&Itemid=84&lang=en.

NBI (2012b). Nile Basin Initiative website. Retrieved from www.nilebasin.org/newsite/.

NBI (2016). *Building on Shared Benefits: Transforming Lives in the Nile Basin*. Nile Basin Initiative, Entebbe, Uganda.

Nile Agreement (1929). Exchange of Notes between Great Britain and Northern Ireland and Egypt in Regard to the Use of the Waters of the River Nile for Irrigation Purposes, Cairo, May 1929. *League of Nations Treaty Series* 2103.

Nile Agreement (1959). Agreement between the Arab Republic of Egypt and the Republic of the Sudan for the Full Utilization of the Nile Waters. 453 *United Nations Treaty Series* 64 (1963).

Nile Cooperative Framework Agreement (2010). Retrieved from http://international water law.org/documents/regionaldocs/Nile_River_Basin_Cooperative_Framework_ 2010.pdf.

Nile Treaty (1902). Treaty between Ethiopia and the United Kingdom, Relative to the Frontiers between the Anglo-Egyptian Sudan, Ethiopia, and Eretria. Retrieved from http://untreaty.un.org/ilc/documentation/english/a_5409.pdf.

Okidi, O. (1994). History of the Nile and Lake Victoria Basins Through Treaties. In P.P. Howell and J.A. Allan (eds), *The Nile: Sharing A Scarce Resource* (pp. 321–350). Cambridge: Cambridge University Press.

Salman, S.M.A. (2001). Legal Regime for Use and Protection of International Watercourses in the Southern African Region: Evolution and Context. *Natural Resources Journal, 41*(4), 981–1022.

Salman, S.M.A. (2003). From Marrakech Through the Hague to Kyoto: Has the Global Debate on Water Reached a Dead End?. *Water International, 28*(4), 491–500.

Salman, S.M.A. (2007a). The United Nations Watercourses Convention Ten Years Later: Why has its Entry into Force Proven Difficult?. *Water International, 32*(1), 1–15.

Salman, S.M.A. (2007b). The Helsinki Rules, the United Nations Watercourses Convention and the Berlin Rules: Perspectives on International Water Law. *International Journal of Water Resources Development, 23*(4), 625–640.

Salman, S.M.A. (2009). *The World Bank Policy for Projects on International Waterways: An Historical and Legal Analysis.* Leiden: Martinus Nijhoff and World Bank.

Salman, S.M.A. (2010). Downstream Riparians can also Harm Upstream Riparians: The Concept of Foreclosure of Future Uses. *Water International, 35*, 350.

Salman, S.M.A. (2011a). The New State of South Sudan and the Hydro-Politics of the Nile Basin. *Water International, 36*(2), 154–166.

Salman, S.M.A. (2011b). Grand Ethiopian Renaissance Dam: Challenges and Opportunities. *The CIP Report, 10*(4), 21–29. George Mason University School of Law. Retrieved from https://cip.gmu.edu/wp-content/uploads/2013/06/111_The-CIP-Report-October-2011_ DamSector.pdf.

Salman, S.M.A. (2012a). Water Resources in the Sudan North–South Peace Process and the Ramifications of the Secession of South Sudan. In J. Troell, E. Weinthal and M. Nakayama (eds), *Water and Post-Conflict Peacebuilding.* Washington DC: Environmental Law Institute, London: Earthscan.

Salman, S.M.A. (2012b). Outcome of the Nairobi Nile Council of Ministers Meeting: An Inevitable Consequence of a Level-Playing Field?. *International Water Law Project Blog.* Retrieved from www.internationalwaterlaw.org/blog/2012/02/14/outcome-of-the-nairobi-nile-council-of-ministers-meeting-%E2%80%93-an-inevitable-consequence-of-a-level-playing-field/.

Salman, S.M.A. (2013). Mediation of International Water Disputes: The Indus, the Jordan and the Nile Basins Interventions. In L. Boisson de Chazourenes, C. Leb and M. Tignino (eds), *International Law and Freshwater: The Multiple Challenges.* Cheltenham: Edward Elgar Publishing.

Salman, S.M.A. (2014). Sudan Continues Relinquishing a Growing Portion of Nile Waters Share. Retrieved from www.salmanmasalman.org/sudan-continues-relinquishing-a-growing-portion-of-nile-water-share/.

Salman, S.M.A. (2015). Entry into Force of the UN Watercourses Convention: Why Should it Matter?. *International Journal of Water Resources Development, 31*(1), 4–16.

Salman, S.M.A. (2016). The Grand Ethiopian Renaissance Dam: The Road to the Declaration of Principles and the Khartoum Document. *Water International, 41*(4), 512–527.

Swain, A. (1997). Ethiopia, the Sudan, and Egypt: The Nile River Dispute. *The Journal of African Modern Studies, 35*(4), 675–694.

Tvedt, T. (2004). *The River Nile in the Age of the British: Political Ecology and the Quest for Economic Power*. London: I.B.Tauris.

Waterbury, J. (2002). *The Nile Basin: National Determinants for Collective Action*. New Haven, CT: Yale University Press.

World Bank (2006). Ethiopia: Managing Water Resources to Maximize Sustainable Growth: A World Bank Water Resources Assistance Strategy for Ethiopia. Washington, DC.

Yihdego, Z. (2017). The Fairness 'Dilemma' in Sharing the Nile Waters: What Lessons from the Grand Ethiopian Renaissance Dam for International Law?. *International Water Law, 2*(2), 1–80.

3 Agreement on declaration of principles on the GERD

Levelling the Nile Basin playing field

Salman M.A. Salman

Introduction

A number of characteristics distinguish the Nile River Basin from other international rivers. The most prominent and widely quoted distinction of the Nile is that it is the longest river in the world, flowing for more than 6,660 kilometres from its origins in the hills of Burundi and Rwanda to the Mediterranean Sea. The river passes, in this long and turbulent journey, through the territories of eleven countries that share, with varying interests, contribution and uses, the waters of the river. The interests of Egypt, Ethiopia, South Sudan and Sudan are classified as very high; those of Uganda as high; with the interests of Burundi, Kenya, Rwanda and Tanzania as moderate; while those of the Democratic Republic of Congo and Eritrea as low (Salman 2013a). About 250 million people in these eleven countries live or depend on the Nile, and the number is expected to reach 500 million by 2050.

The size of its lakes and swamps is another characteristic of the Nile River. One of the two main tributaries of the Nile, the White Nile, originates in Lake Victoria, the largest in the developing world, and the second in size in the world, after Lake Superior in North America. The White Nile struggles through its journey in South Sudan with the most extensive swamps in the world, the Sudd that could reach 30,000 square kilometres in size in some years (Collins 1996: 11–12).

A third distinguishing factor for the Nile is the large number of agreements concluded, but not recognised by one or more of the parties (Garretson 1967). The 1902 Agreement between Britain and Ethiopia, and the 1929 Agreement between Britain and Egypt, are the most widely disputed of the Nile agreements. Egypt claims that both agreements grant Egypt veto power over projects on the Nile in other Nile states, if Egypt decides that any such project would cause harm to its Nile water interests. Ethiopia rejects the 1902 Agreement (Degefu 2003), while Kenya, Tanzania and Uganda declared in 1962, under the Nyerere Doctrine, that they were no longer bound by the 1929 Agreement (Mekonnen 1984). Egypt, on the other hand, claims that the two agreements are valid and binding on those countries.

A third agreement that is resented by the other Nile riparians is the 1959 Nile Waters Agreement between Egypt and Sudan. The two countries have determined that the entire flow of the Nile, measured at Aswan, comprises 84 billion cubic metres (BCM). After deducting the evaporation losses at the Aswan High Dam reservoir of 10 BCM, the two countries divided the remaining 74 BCM between them – 55.5 BCM for Egypt and 18.5 BCM for Sudan – leaving nothing for the other riparian states. Worse, the 1959 Agreement stipulates that since other riparian countries of the Nile may claim a share in the Nile waters, both countries agree to study together these claims and adopt a unified view thereon. If such studies result in the possibility of allotting an amount of the Nile waters to one or the other of these states, then the value of this amount, as measured at Aswan, shall be deducted in equal shares from the share of each of the two countries. The Permanent Joint Technical Committee established by the two countries under the 1959 Agreement shall make arrangements with the concerned authorities in other countries in connection with the control and checking of the agreed amounts of Nile waters so allotted. Thus, the two countries granted themselves the right to determine if any of the other Nile riparians is entitled to any share of the Nile waters, and how much this share should be; and have included in the Agreement arrangements to ensure that the allotted amount by them is not exceeded by that state.

These disputed agreements complicate the Nile contour further when they get intertwined with another basic characteristic of the Nile – the large number of existing and planned dams on the Nile River and its main tributaries. Some of these dams are among the oldest and largest in the world (Waterbury 2002). And the most recent and newest, the Grand Ethiopian Renaissance Dam (GERD), has proven to be the most disputed dam on the river and the one that has drastically changed the paradigm and legal contours of the Nile, as is discussed in detail in this chapter.

Dams and the Nile troubled waters

The history of dam building on the Nile dates back to the nineteenth century. Construction of the Aswan Dam by the British in Egypt commenced in 1894 and was completed in 1902, as the first dam on the Nile River and the largest in the world at that time. Its height was increased twice, in 1908, and then in 1933, so as to increase its storage capacity to double the original one. Egypt also constructed a series of barrages – the Essna, Naga Hammad, Assuit and Edfina – to divert the Nile waters to expand its irrigated areas and water use.

In 1960, following the conclusion of the 1959 Nile Waters Agreement with the Sudan, Egypt started constructing the Aswan High Dam (AHD) that was completed in July 1970, and officially inaugurated in January 1971 as the largest dam in the world at that time, with a reservoir capacity of 162 BCM. The AHD was built by Soviet funds and technology, after the West and the World Bank withdrew their offer to finance the dam. That situation heightened the scramble of the superpowers over the Nile Basin.

Sudan's dam construction activities started in 1919 with the building on the Blue Nile of the Sennar Dam that was completed in 1925 to irrigate the Gezira

Scheme. However, to get Egypt's consent to this dam, the Anglo-Egyptian administration in the Sudan had to agree to the construction of the Jebel Aulia Dam on the White Nile in the Sudan, for the exclusive uses and benefits of Egypt. It was the first dam ever to be built in the territory of one nation for the exclusive use of another nation. The Jebel Aulia Dam stored annually 3.5 BCM of the waters of the White Nile during the flood season of the Blue Nile, to be released for the use in Egypt when the waters of the Blue Nile diminish to their low level. The construction of the Jebel Aulia Dam started in 1933 and was completed in 1937, and it was the first dam on the White Nile. The dam lost its importance to Egypt when the AHD, with its huge storage capacity of 162 BCM, was completed and inaugurated in 1971, and the Jebel Aulia Dam was returned to the Sudan six years later, in 1977 (Collins 2002). The Jebel Aulia Dam is a major burden on the Nile waters and the Sudan due to the heavy annual evaporation of 2.5 BCM, and the limited uses Sudan currently makes out of the dam.

The formula of 'a dam for a dam' continued and was replicated under the 1959 Nile Waters Agreement between Egypt and Sudan. That Agreement allowed Sudan to build the Roseires Dam on the Blue Nile to irrigate the extension of the Gezira Scheme (Managil extension), after Sudan agreed to the construction of the AHD. The AHD flooded large areas of Sudan and forced the relocation 50,000 Sudanese Nubians, whose lands and main town, Wadi Halfa, and twenty-seven of their villages, were submerged under the reservoir of the AHD that extended for 150 kilometres inside Sudan (Dafalla 1975). The Roseires Dam was built, similar to the Sennar Dam, on the Blue Nile, and was commissioned in 1966. Sudan also built the Khashm el Girba Dam to irrigate the New Halfa Agricultural Scheme where the Sudanese Nubians affected by the construction of AHD, were forcibly relocated (Abdalla 2000). The Khashm el Girba Dam, which was completed in 1964, was the first dam to be constructed on the Atbara River, the last of the tributaries of the Nile River in its journey northwards.

The most recent of the Sudanese dams included the Merowe Dam on the Nile River itself, at the fourth cataract. It was completed in 2009 for the generation of hydropower. Plans for an irrigation scheme, although included as a part of the project design, have not yet been put into effect. In January 2017, Sudan completed and inaugurated the Upper Atbara and Setit Dams, basically to replace the Khashm el Girba Dam that has been negatively affected by the huge accumulation of sediments. The purposes of these two dams include irrigation, drinking water supply and the generation of hydropower.

Sudan has plans to build three more dams – Kajbar, Shereik and Dal – all on the Nile River for generation of hydropower and irrigation. However, questions persist about the source of funding for these dams, despite the Saudi preliminary offer to finance these dams. There is also the fierce opposition of the affected communities in the three locations, fearing the fate of those forcibly relocated as a result of both the AHD and the Merowe Dam. Egypt is concerned about the irrigation plans of the Merowe Dam, as well as the expansion of the uses of the planned three dams beyond generation of hydropower, to include irrigation schemes.

Uganda started its dam-building activities in the late 1940s with the Owen Falls Dam that was built after intensive negotiations with Egypt, resulting in a number of agreements between Britain and Egypt (Okidi 1994). The dam was built on the Victoria Nile, immediately after its exit from Lake Victoria, and was completed in 1953 for the generation of hydropower. This project was rehabilitated in the 1980s, and renamed Nalubaali. It was followed by the Kiira Dam projects, also on the Victoria Nile for generation of hydropower. In 2007, Uganda was able to secure the funding from the World Bank Group for the Bujagali Dam that was completed in 2012 for generation of hydropower. Uganda also started, with Chinese funding and technology, constructing the Isimba dam in 2014, and has completed the studies for the Karuma Dam, while the studies for the Kalagala Dam are ongoing. All these dams are on the White Nile, known in Uganda as the Victoria Nile in the first stretches, and as the Albert Nile in the second and last stretches inside Uganda, before the river crosses into South Sudan. Egypt is also concerned about the fast-moving dam-building activities of Uganda, particularly with its absence in overseeing these dams, as happened with the Owen Falls Dam.

Ethiopia was the last of the Nile Basin countries to enter the dam-building competition. Ethiopia started with the Fincha Dam – a small dam on the Blue Nile that was completed in 1973 for the generation of hydropower. It was followed by two small dams, the first and second Tis Abbay projects, on Lake Tana for the generation of hydropower also. However, at the beginning of this century, Ethiopia moved to the fourth project on the Nile, the largest and most significant of all. That was the Tekeze Dam on the Atbara River, which was started in 2002 and was completed in 2010. This large dam, with a height of 188 metres and storage capacity of about 10 BCM, generates more than 360 megawatts of electricity. Concurrently, Ethiopia also built the Tana–Beles project that connected the Beles River (a tributary of the Blue Nile) with Lake Tana, and generated hydropower at that connecting point. The project was also completed in 2010. The last two dams generated the incentives and appetite for Ethiopia to push ahead with the construction of the GERD, as discussed in the next part of this chapter. These two projects also presented and resulted in Ethiopia's practical challenge to the 1902 Agreement referred to earlier. Ethiopia is also planning on building a number of dams, including Karadobi, Mandaya, Didessa and Mabil, all on the Blue Nile (World Bank 2006).

Dam-building activities by Ethiopia on other rivers, particularly the Omo River, are going on concurrently with the dams on the Nile (Salman 2011). Thus far, Ethiopia has completed three dams for the generation of hydropower on the Omo River, the fourth one is under construction, and the fifth is being studied. The Omo River flows into Lake Turkana (shared by Ethiopia and Kenya), and Ethiopia's dam activities on the river are raising concerns about the environmental well-being of the lake, as well as livelihoods of the Turkana indigenous peoples, who live and depend on the lake.

The Grand Ethiopian Renaissance Dam

Ethiopia moved its ambitious dam-building programme to the Grand Ethiopian Renaissance Dam (GERD) in 2011. The genesis of the GERD can be traced to the studies conducted by the United States Bureau of Reclamation in the 1960s (USBR 1964), following, and as a result of, the commencement of the construction of the AHD in Egypt by the Soviet Union. The studies indicated that the hydropower potential of Ethiopia exceeds 45,000 megawatts, with 30,000 from the Nile. The studies recommended a number of dam projects on the Nile, including the Border Dam on the Blue Nile, close to the boundaries with Sudan. The study for this dam was later updated, and in March 2011, Ethiopia announced its plans to construct a dam on the Blue Nile, about 20 kilometres from the Sudanese borders. This dam became known as the Millennium Dam, and the name was later changed to the Grand Ethiopian Renaissance Dam (GERD).

The construction of the GERD commenced at the beginning of April 2011, a few days after the announcement was made, and when Egypt was completely immersed in its January 2011 revolution that toppled President Hosni Mubarak in February of that year. The GERD is a large dam, with a height of 145 metres and a storage capacity of 74 BCM – as opposed to the original Border Dam, which had a storage capacity of 14 BCM. The reservoir is more than twice the size of Lake Tana, with a length of 245 kilometres, and is expected to cover an area of 1,874 square kilometres. The GERD is supplemented by a saddle dam with a length of 5 kilometres and a height of 245 metres. The initial installed capacity of the GERD is 6,000 megawatts, with sixteen turbines, each expected to generate 375 megawatts. However, Ethiopia indicated in 25 February 2017, that the installed capacity of the GERD has been revised upwards from 6,000 to 6,450 megawatts 'as a result of the improvement made on generators to boost the capacity of the power plant' (Ezega 2017). This is more than three times the electricity being generated by the AHD of 2,100 megawatts.

Table 3.1 The Aswan High Dam, the Merowe Dam and the Grand Ethiopian Renaissance Dam (GERD)

Name of dam	Aswan High Dam	Merowe Dam	GERD
Year of completion	1970	2009	2017 (expected)
Height (metres)	183	67	145
Storage (BCM)	162	12	74
Size of reservoir (km^3)	6,000	800	1,800
Length of reservoir (km)	500	176	245
Electricity generated (installed capacity)	2,100	1,250	6,450
Evaporation (BCM)	10	2	2
No. of people forcibly relocated	120,000	50,000	14,000

Source: Salman 2016.

The current estimates of the cost of the GERD are about $4.8 billion. Ethiopia announced from the start that it intends to finance the GERD from its own resources, and through issuance of bonds to Ethiopians and interested foreign nationals, with no resort to foreign borrowing. The GERD is being constructed by the Italian company, Salini Impregilo, and will be, upon completion in 2017, the largest in Africa, and the tenth largest in the world. Ethiopia has stated repeatedly that the dam's sole purpose is the generation of hydropower and is not intended for irrigation purposes, since the terrain of the region does not allow for the development of any irrigated agriculture.

There is no doubt that this is a huge project when compared with the other two existing largest dams in Egypt and Sudan – the Aswan High Dam and the Merowe Dam. The table below compares the AHD, the Merowe Dam and the GERD:

Initial Egyptian and Sudanese reaction and the road to accepting the GERD

Egypt and Sudan reacted immediately after the Ethiopian announcement was made, protesting and denouncing the Ethiopian decision, and declaring their strong opposition to the GERD. Both countries contended that the GERD would considerably decrease the amount of Nile waters reaching Sudan and flowing thereafter to Egypt. Egypt further claimed that the GERD would turn large tracts of its irrigated lands into desert,[1] and would result in a considerable decrease of the hydropower generated by the AHD. Both countries demanded that the different studies for the GERD be provided to them, as a part of the notification process, so that they can assess the harm that the GERD would cause each of them. Sudan was also concerned about the safety of the GERD, and the extensive harm Sudan could suffer, if the GERD were to crack, fail or collapse.

However, the Sudanese position began to witness some gradual, but steady changes, and a large number of water experts started arguing that Sudan would actually benefit from the GERD. They explained that the benefits to Sudan include the entrapment of the huge sediments that the Blue Nile carries and brings annually to Sudan, which have caused the Sennar and Roseires Dams to lose more than half their storage and electricity generating capacity. They also include the regulation of the flow of the waters of the river, thus putting an end to the recurrent flooding and the destruction to property and crops, caused by the seasonal flow of the Blue Nile. Regulation of the flow, it is further argued, would help Sudan to increase its crop rotations to two, or even three a year, from the current single rotation, due to the seasonality of the flow of the waters of the river. This would also help the replenishment of groundwater in the surrounding areas throughout the year (Salman 2013a).

It should be added that Sudan is one of the countries with the lowest amount of water stored (about 10 BCM in its five dams, compared to 162 BCM for Egypt in AHD). Sudan could also negotiate the storage of some of its Nile waters in the GERD reservoir (Salman 2011). It is to be noted that Sudan uses only

12 BCM of the 18.5 BCM allotted to it under the 1959 Nile Waters Agreement with Egypt. Sudan's unused portion of its share of 6.5 BCM has been crossing the northern Sudanese borders to Egypt annually and regularly since 1959 (Salman 2014). The sale of the GERD electricity to Sudan, for which the power connection lines have already been completed and are operational, as well as to Egypt, has also been cited as one major additional benefit to both countries which face a huge deficit in electric power.[2] Indeed, Sudan already purchases 250 megawatts of electricity from Ethiopia through these power lines. The power connection lines with Egypt are underway.[3] It should be noted, and indeed underscored, that the Ethiopian electricity is cheaper than that generated in either Sudan or Egypt.[4]

Those arguing in favour of the GERD for Sudan have pointed out that the only two issues of concern to Sudan should be the safety of the dam and the length of the period during which the reservoir will be filled.

Despite the Sudanese position favouring the dam, or perhaps because of that position, the Egyptian opposition to the GERD continued to mount. Egypt continued to demand that Ethiopia formally notifies both countries of the project, provides them with all available information, and allows time for their response before it moves on with the construction of the GERD. Ethiopia has always rejected the requests for notification for any of its Nile projects, claiming that the hidden objective of the Egyptian demand is for Egypt, after it is notified, to invoke, and claim that the 1902 Agreement is valid and binding on Ethiopia. Furthermore, Ethiopia contends that Egypt and Sudan never notified Ethiopia of any of their projects on the Nile, accordingly, and as a matter of reciprocity, Ethiopia is under no obligation to notify either of them.

As the impasse persisted, Egypt and Sudan agreed with the Ethiopian proposal in September 2011 of establishing an International Panel of Experts, consisting of ten members, two from each of the three countries, and four from outside the Nile Basin countries. The terms of reference of the Panel included identifying any negative impacts of the GERD on Sudan and Egypt, and recommending ways of mitigating such impacts. The Panel was established in November 2011 and commenced its work soon after that.

Ethiopia's construction of the GERD continued unabated, and on 28 May 2013, Ethiopia diverted the Blue Nile from its main and natural course of flow so it could commence construction of the GERD. Egypt vehemently protested the diversion of the Blue Nile and stated that Ethiopia should have waited for the issuance of the Panel's report before making any drastic decisions on the GERD. In the meeting that took place in early June 2013, between the Egyptian government and the opposition parties to discuss the diversion, some participants called for a military attack on the GERD, as well as arming and funding opposition groups in Ethiopia to carry out that task. Unbeknown to those attending, the meeting was broadcast live (Al Arabiya 2013). That resulted in a major international embarrassment to Egypt.

Three days after the diversion of the Blue Nile by Ethiopia, the International Panel of Experts issued its report (International Panel of Experts 2013). The report, dated 31 May 2013, was signed by all the ten Panel members, and recommended the carrying out of more in-depth studies on the effects of the GERD on Egypt and Sudan. Consequently, Egypt demanded that construction of the GERD be suspended until the studies were completed. Ethiopia, on the other hand, claimed that the Panel did not recommend such suspension, and the construction of the GERD and the carrying out of the studies could, and would, go concomitantly. Another impasse ensued, but the three parties continued to talk to each other, and agreed to meet at the ministers of water resources level to discuss their differences.

The first such tripartite ministerial level meeting took place on 4 November 2013 in Khartoum. The meeting started with some optimism that the parties would be able to resolve the pending issues. However, that optimism was soon dashed by the strong tide of the Blue Nile, when Egypt insisted on its demand of the suspension of the work on the GERD until the studies had been completed. Ethiopia rejected that demand. Another major difference erupted during that meeting. Egypt demanded that the studies be carried out by international independent experts, while Ethiopia insisted that the role of international experts had ended with the submission of the Panel report, and that the studies would be carried out by Ethiopian experts. Thus, the first ministerial meeting ended in failure, but the three parties agreed to meet again in December 2013.

However, the Nile alliance of Egypt and Sudan against the rest of the Nile Basin countries, that was born when the two countries concluded the 1959 Nile Waters Agreement, would soon be strongly shaken. Sudan announced on 4 December 2013, through its president of the Republic himself, and in the presence of the Ethiopian prime minister, the support of Sudan to the GERD (Eleiba 2013). The announcement was made during the inauguration of the power connection lines between Ethiopia and Sudan, which were completed that month. Indeed, it was announced during the event that Ethiopia had started selling Sudan 100 megawatts of electricity, to be carried through the newly inaugurated power lines. The support of Sudan for the GERD was no doubt a ground-shaking decision, which virtually ended its more than half a century Nile alliance with Egypt, and left Egypt out in the cold.

The announcement angered Egypt and resulted in a more defiant Egyptian attitude towards the two new allies, and the GERD, when the second tripartite ministerial meeting took place two days later. Egypt continued to insist on its two demands of the suspension of construction of the GERD until the studies had been completed, and that the studies should be carried out by independent international experts. Ethiopia showed some willingness to compromise. The compromise presented by Ethiopia was that the studies could be carried out by experts from the three countries, instead of just Ethiopian experts. However, that concession was not accepted by Egypt, and the December meeting ended in a failure to reach an agreement on either of those two issues. Despite this failure, the parties agreed to meet again in a month's time.

The third tripartite ministerial meeting took place in January 2014 in Khartoum. However, Egypt and Ethiopia insisted on their respective positions, and the third meeting ended with no agreement on these two hotly disputed matters. Worse, no discussion on the time and place of a fourth ministerial meeting took place – the parties departed Khartoum and stopped talking to each other.

Ethiopia continued the construction of the GERD, betting that time was on its side, and every day that passed would help in making the GERD a fait accompli. On the other hand, Egypt's bet was that the cost of constructing the GERD – close to $5 billion – was far beyond Ethiopia's financial means, and that work would stop, sooner or later, as Ethiopia would not be able to raise the necessary funds in the international money markets or from any of the international financial institutions. Sudan's bet was that the two countries would continue to need a third party to talk through to each other, and that despite its support of the GERD, Sudan would continue to be that third party. Evidently, Ethiopia's bet has continued to have the best chances of winning.

Five more months were to elapse before the stagnant waters of negotiations on the GERD would start moving again. The African Union meeting in Malabo, Equatorial Guinea, in June 2014, provided an opportunity for the new President of Egypt, Field Marshall Abdel Fattah El-Sisi, and the Ethiopian Prime Minister Hailemariam Desalegn to meet, for the first time, face to face without a third-party intervention. Surprisingly, the meeting went well and the two leaders agreed on general principles of cooperation in a number of fields, including the Nile waters and the GERD. It was further agreed that the tripartite meetings should resume soon.

And in fact, the meetings did resume on 25 and 26 August 2014 in Khartoum. The Press Release on the fourth tripartite meeting of the ministers of water resources echoed the spirit of the Malabo meeting, and included a compromise on the two pending issues. The three parties agreed that the two studies recommended by the Panel would be undertaken by international consultants under the supervision of a national panel, consisting of four members from each of the three countries. There was no mention of the demand of Egypt for suspension of the work on the GERD while the studies were being undertaken. Thus, Egypt dropped its heavy demand regarding suspension of the work on the GERD, while Ethiopia agreed to the participation of international experts in the carrying out of the studies. The compromise was face-saving for each of the two parties, but the clear signal that emerged by that time was the Egyptian acceptance of the fact that the GERD has indeed become a reality, and that the GERD construction would continue unabated.

The tripartite ministerial meetings continued, and the fifth one was held in Addis Ababa on 22 and 23 September 2014. That was the first tripartite meeting to take place outside Khartoum, which is conveniently located midway between Cairo and Addis Ababa. Khartoum must have started to get worried that it was about to lose that strategic position. The parties agreed that the international consultants would be selected by the national experts from the three countries, who together were called the Tripartite National Committee (TNC).[5] It was further agreed that the TNC would have its chairmanship, as well as its meetings, by rotation among the three countries.

During the fifth tripartite meeting, a noteworthy development took place. The three ministers visited the GERD location. And although the Egyptian minister indicated that the visit was a technical and not a political visit, there was definitely a big sigh of relief in Addis Ababa that Egypt could no longer deny the fact that it has indeed accepted the GERD.

Improvements in the negotiations environment continued, and the sixth tripartite meeting took place in Cairo on 16 October 2014. Furthermore, the three ministers met with President El-Sisi, and the discussions went beyond the GERD and touched on other pending Nile issues. At the end of the meeting, the ministers agreed that the international consultants would be selected, and the studies be completed, within six months.

Meanwhile, another significant development regarding the GERD took place at the international level. In November 2014, the Massachusetts Institute of Technology (MIT) issued a report prepared by seventeen international water experts, none of them from the Nile Basin countries (MIT 2014). The report indicated that the experts: 'support the Ethiopian strategy for developing its water resources in the Blue Nile Basin, and acknowledge that the GERD is the first major step in the implementation of this economic development strategy' (MIT 2014). Furthermore, the report communicated the thoughts of the experts, who prepared it on four issues:

1 Need for an agreement on the coordinated operation of the GERD with the Aswan High Dam (AHD),
2 Technical issues regarding the design of the GERD,
3 Need for an agreement on the sale of hydropower from the GERD, and
4 Potential downstream impact on Egypt and Sudan, particularly in agriculture.

(MIT 2014)

The report went on to indicate that:

It is important to emphasize that, in making these assessments, we did not have access to some of the relevant information about the GERD. Thus, some of our concerns may be assuaged when Ethiopia makes more information about the GERD publicly available to the international community.

(MIT 2014)

This report has no doubt added a considerable weight to Ethiopia's position, and strengthened its resolve to go ahead with the GERD as planned, now that a reputable academic engineering institution has endorsed its position and right to utilise its water resources for development. It may even be argued that the MIT report has in fact influenced the outcome of the seventh tripartite meeting.

In addition, most of the African countries supported the GERD, or just kept quiet. European countries were competing on selling the electrical equipment and turbines for the GERD and other hydropower projects in Ethiopia. In fact, even some of the Arab countries, including Saudi Arabia and Qatar, showed signs of supporting publicly and explicitly Ethiopia and the GERD (Al-Youm 2017; Daily News Egypt 2016).[6] The live broadcast of the meeting in early June 2013 of the Egyptian government and opposition parties, where military options against the GERD were discussed, continued to be a major international embarrassment to Egypt. These developments must have pushed Egypt into the road of accepting the GERD, albeit reluctantly. They must have also given Sudan some comfort with regards to its decision of supporting the GERD.

The seventh tripartite meeting took place in Khartoum from 3 to 5 March 2015. The meeting was attended by both the Ministers of Water Resources in the three countries, as well as the three Ministers of Foreign Affairs. Participation of the six ministers indicated the agreement of the three parties to have both technical as well as political negotiations on the GERD, and the desire of the parties to reach an agreement. Indeed, the six ministers announced at the end of the meeting that they have reached an agreement on the GERD, and the agreement was being reviewed by the two presidents of Egypt and Sudan, as well as the prime minister of Ethiopia.

The Agreement is titled 'Agreement on Declaration of Principles between the Arab Republic of Egypt, the Federal Democratic Republic of Ethiopia, and the Republic of the Sudan on the Grand Ethiopian Renaissance Dam Project (GERDP)' (DoP 2015). It was signed in Khartoum on 23 March 2015, by the two Presidents, Abdel Fattah El-Sisi and Omar Hassan Ahmad al-Bashir, and Prime Minister Hailemariam Desalegn themselves, despite the presence of the Ministers of Foreign Affairs, as well as those of the Water Resources of the three countries, in the ceremony.

Agreement on Declaration of Principles on the GERD

The Agreement on Declaration of Principles (DoP) on the GERD consists of a preamble and ten principles, four of which relate to the GERD, while the other six deal with some basic principles of international water law. The preamble confirmed the significance of the Nile River as a source of livelihood and development for the people of the three countries, thus restating a basic principle of international water law of equality of all the riparians in the sharing and uses of the common river.[7]

Article 1 of the DoP dealt with the principle of cooperation, based on common understanding, mutual benefits, good faith and the principles of international law, as well as understanding upstream and downstream water needs in its various aspects. Article 2 stipulated clearly the recognition of Egypt and Sudan of the purpose of the GERD as power generation, contribution to economic development, promotion of transboundary cooperation and regional integration through generation of reliable and sustainable energy.

Article 3 obliged the three parties to take all appropriate measures to prevent the causing of significant harm. In line with the United Nations Watercourses

Convention provisions, Article 3 went on to state that where significant harm nevertheless is caused, the state whose use causes such harm shall, in the absence of agreement to such use, take all appropriate measures, in consultation with the affected state, to eliminate or mitigate such harm and, where appropriate, to discuss the question of compensation.[8]

Article 4 dealt with the principle of equitable and reasonable utilisation, and laid down the same factors set forth in the Watercourses Convention, but added two more factors dealing with the contribution of each state to the waters of the Nile River, and the extent and proportion of the drainage area in the territory of each basin state. These two factors are copied from the Nile Basin Cooperative Framework Agreement, which Egypt and Sudan oppose, and which Ethiopia has signed and ratified.[9] The two factors certainly weigh in favour of Ethiopia, which contributes about 86 per cent of the Nile waters, and is third, after Sudan and South Sudan, in the size of the Nile drainage area in its territories (Salman 2016).

Article 5 dealt with the principle to cooperate in the first filling and operation of the dam. It called for the implementation of the recommendations of the International Panel of Experts, and for respect of the outcomes of the Tripartite National Committee (TNC) final report on the joint studies recommended by the Panel throughout the different stages of the dam project (Zhang, Erkyihun and Block 2016; Wheeler *et al.* 2016). Article 5 also asked the three countries to utilise the final outcomes of the studies to agree on the guidelines and rules for the first filling of the GERD. However, the Article clarified that this would take place 'in parallel with the construction of the GERD'. Further, Article 5 asked the three parties to agree on guidelines and rules for the annual operation of the GERD, but subjected this to adjustments, which the owner of the dam may take from time to time. It then required Ethiopia to inform the two other countries of any unforeseen circumstances requiring adjustments in the operation of the GERD.

Article 5 also addressed the issue of coordination of the operation of the GERD with the downstream reservoirs, and asked the three countries to set up, through the line ministries, appropriate coordination mechanisms among them within fifteen months from the inception of the studies recommended by the Panel.[10]

Article 6 dealt with the issue of confidence building and stipulated that priority would be given to the downstream countries in the purchase of power generated by the GERD. Article 7 addressed the issue of exchange of data and information between the three parties for carrying out the studies recommended by the Panel, and to be carried out by the TNC.

Article 8 dealt with the issue of the dam safety. The Article underscored the appreciation of the three countries to the efforts undertaken thus far by Ethiopia in implementing the recommendations of the Panel with regards to the dam safety, and called on Ethiopia to continue to implement in good faith these recommendations.

Article 9 returned to the concept of cooperation, and required the three parties to cooperate on the basis of sovereign equality, territorial integrity, mutual benefit and good faith, in order to attain the optimal utilisation and adequate protection of the Nile River (Tawfik 2016; Cascão and Nicol 2016). This is a clear reiteration of the concept of equality of all the riparians stipulated in Article 1 of the Agreement, discussed above.

Article 10, the last article of the DoP, dealt with the principle of peaceful settlement of disputes. It required the three countries to settle any disputes arising out of the interpretation or implementation of the agreement amicably through consultation or negotiation in accordance with the principle of good faith. Failing that, the parties may jointly request conciliation or mediation, or refer the matter for consideration of the heads of states/governments. Thus, the DoP does not include a resort to arbitration or to the International Court of Justice.

This is no doubt a landmark development in the history of the Nile Basin. Egypt and Sudan, which divided the entire flow of the Nile between them under the 1959 Agreement, and asked the other riparians to apply to them for any share of the Nile waters, have finally recognised the equality of all the Nile states. They have also acknowledged the right of these states to utilise the waters of the Nile River for the sustainable development of their people. The playing field has clearly, and finally, been levelled for the first time in the history of the Nile Basin. Furthermore, the DoP would soon be confirmed through another important document, as discussed below.

Challenges faced with the panel-recommended studies

As discussed above, the three parties agreed that two studies, one on hydrological remodelling and the other on the impact of the GERD on Sudan and Egypt, which were recommended by the International Panel of Experts, would be carried out. The compromise modus operandi for carrying out the studies, agreed after lengthy discussions, was that international consultants would carry the studies under the overall supervision of the Tripartite National Committee (TNC). The TNC recommended the French firm, BRLi Group, and the Dutch firm, Deltares, to undertake the studies; and the recommendation was endorsed by the subsequent tripartite ministerial meeting. Apparently, the three parties agreed that each of Ethiopia and Egypt would choose one firm, and Sudan would provide the endorsement for the two firms.

However, differences erupted among the three parties on the detailed terms of reference of each of the two firms, as well as the role of each of the two firms in the carrying out of the studies. The differences soon moved to the two firms themselves. Deltares rejected the secondary role assigned to it, and demanded an equal role with BRLi in the carrying out of the studies. The issue occupied the meetings of the TNC, and the two tripartite meetings of the six ministers that took place in May and August 2015. In September 2015, Deltares announced that it was withdrawing, and would not participate in the carrying out of the two technical studies.

The TNC continued to meet to try to address the issues and differences related to the two studies, and reach a compromise thereon. A third tripartite meeting of the six ministers took place in Khartoum on 11 and 12 December 2015, amid reports that the studies were facing a number of difficulties, and amid rumours that Egypt was about to withdraw from the Agreement on Declaration of Principles on the GERDP concluded on 23 March 2015. The third tripartite meeting discussed these issues, and it was agreed that a fourth tripartite meeting for the six ministers would take place in Khartoum on 27 and 28 December 2015.

As preparations were going ahead for the fourth tripartite meeting, Ethiopia announced on 25 December 2015, that it had returned the Blue Nile to its natural and main course of flow after it had completed construction of the cement and steel work of the GERD. It should be recalled that Ethiopia had diverted the Blue Nile on 28 May 2013, so as to start constructing the GERD. Thirty-one months later, the work was completed, and the river was returned to its natural course. That announcement indicated that the GERD had become, for all practical purposes, a reality. Instead of complicating the fourth tripartite meeting, it could be argued that this development in fact pushed the parties in the direction of another historic agreement, as discussed below. The development confirmed to Egypt that the GERD was now a reality that it could not escape dealing with.

Summary and the outcomes of the fourth tripartite meeting – the Khartoum Document

The fourth tripartite meeting of the ministers of water resources and the ministers of foreign affairs of the three countries took place, as planned, on 27 and 28 December 2015 in Khartoum, about two weeks after the third tripartite meeting. This was the first time that two tripartite meetings would take place in one and the same month. After two days of intensive discussions, the six ministers reached an agreement on the pending issues and signed a document titled 'Summary and the Outcomes of the Meeting' (Khartoum Document 2015).

The Khartoum Document consisted of six provisions and three annexes. The Document confirmed 'the sincere and full commitment of the three countries to adhere to the Agreement on Declaration of Principles (DoP)', putting an end to the rumours about Egypt's imminent move to withdraw from the DoP.

The first paragraph of the Document recorded the endorsement of the six ministers of the choice by TNC of the French firm Artelia to replace the Dutch firm Deltares. Thus, the two French firms – BRLi Group and Artelia – are now the firms that will carry out the two studies. The document also endorsed the selection of the British Law firm of Corbett and Co. to draft the contracts with the two firms and supervise implementation of the legal obligations under the contracts. The three parties agreed that the cost of the three firms will be shared equally by the three countries.

Paragraph 2 of the Document recorded the request by Egypt regarding the addition of two bottom outlets in the main dam, submitted in the previous tripartite meeting, and Ethiopia's written clarification thereon based on the studies of the Dam. After discussions and elaborations on the issue in broad terms, it was agreed to assign a technical team, from the three countries, to meet in Addis Ababa on 3 and 4 January 2016, to discuss the subject and submit a report to the next coming ministerial meeting. The technical team travelled to Ethiopia and it was reported on 8 January 2016 that Ethiopia indicated that it did not agree with the Egyptian request, and thus would not increase the number of the outlets.

Article 3 of the Document reiterated the full commitment of the three countries to implement the provisions of the Agreement on DoP signed in Khartoum on the 23 March 2015. Ethiopia also reiterated its commitment to implement Article 5 of the DoP on the coordination of the operation of the GERD with the downstream reservoirs.

Article 4 of the Document stated that the tripartite joint ministerial meeting would continue to be convened on a regular basis at the same level of participation of the ministers of foreign and water affairs, so as to ensure the timely completion of the two studies, further enhance the positive atmosphere and closely monitor the progress thereon.

Article 5 of the Document dealt with confidence building and stated that the three countries agree to support and encourage efforts aimed at promoting confidence-building measures so as to enhance people-to-people relations of the three countries. Accordingly, Ethiopia invited parliamentarians, the media and public diplomacy groups, of both Egypt and Sudan, to visit the GERD.[11]

Annex A of the Khartoum Document included the names of those who attended the fourth tripartite meeting, while Annex B included the agenda for the meeting. Annex C of the Document dealt with the roadmap for carrying out the two studies, and set forth certain dates for completion of actions thereon. The signature of the contracts with the two French firms and the launching of the studies would take place, as per this Annex, in Khartoum on 1 February 2016.

Despite this clear stipulation, members of the TNC, and the three ministers of water resources did not meet until 7 February 2016, and the contracts had not been signed by that time. The meeting took place in Khartoum and lasted until 11 February. The ministers of foreign affairs did not attend. The Press Release issued on the last day of the meeting indicated, that: 'the TNC deliberated on the clarification issues on the updated technical proposal, financial proposal of the consultant, draft contract document, and agreed to resolve the pending issues through communications among the TNC, the consultant and the legal adviser' (Press Release 2016). The Press Release added that the TNC agreed to hold its eleventh meeting in Addis Ababa, Ethiopia to sign the contract for the two studies based on the completion of the pending issues to be resolved through communication, but the Press Release did not specify the date of the eleventh meeting, or the new date for signature of the contract.

After a series of meetings of the TNC, the contracts were finally signed with BRLi Group and Artelia on 20 September 2016, more than seven months from the originally agreed-upon date. The contract stipulated that the two studies on 'Water Resources/Hydropower System Simulation Model' and 'Transboundary

Environmental and Socio-economic Impact Assessment' would be completed within eleven months, that is, by August 2017.

Thus, the Agreement on Declaration of Principles on the GERD received a major boost, and has been reconfirmed by the Khartoum Document. The Document resolved the dispute regarding the role of the two firms, and added a law firm to oversee compliance with the contracts. With the signing of the contract with the two firms, the GERD seems to have firmly settled in its location on the Blue Nile after the river was returned to its natural course a few days before the signature of the Khartoum Document. By the time the two studies are due for completion in October 2017, it is most likely that the GERD will have started generating its hydropower, at least through the two bottom turbines.

Conclusion

Dam construction on the Nile is now more than 120 years old, with the river dotted by a huge number of existing dams, some under construction and many others planned. Sadly, the scramble for the use of the Nile waters is country-driven, and all the dams are unilateral projects, forgoing the extensive opportunities and benefits of joint, regional and basin wide projects (Blackmore and Whittington 2008).

The Grand Ethiopian Renaissance Dam is another clear major manifestation of this situation and trend. Indeed, in hindsight, with some cooperation and collective action, the GERD could have easily been the replacement of the Aswan High Dam, the Roseires Dam and the Merowe Dam – all the three dams together. The benefits generated from these three dams (hydropower, irrigation water and flood control) could have easily been reaped in the 1960s from a jointly owned and operated GERD (or a larger Border Dam). Egypt and Sudan could have avoided the loss of huge fertile lands estimated to be more than half a million feddans[12] and the forced relocation of more than 120,000 Nubians, in both countries. This joint action could have also saved the large amount of water of 10 BCM lost annually to evaporation in the extensive reservoir of the AHD in the exceptionally hot deserts of Egypt and northern Sudan, as well as the historical archaeological sites that have been submerged under the huge AHD reservoir, particularly in Sudan.

Similarly, in 2011, the GERD could have been a tripartite project, funded, owned and operated by the three countries. In addition to sharing the extensive benefits of the GERD, the concerns of Egypt and Sudan (period of filling the reservoir and dam safety) could have been adequately and appropriately addressed through the joint operation, and the resulting involvement of the international financial institutions. Yet, this second opportunity was also lost when Egypt and Sudan ignored the early Ethiopian proposal of such joint ownership. Absence of cooperation and the narrowly, and poorly, perceived national interests aborted that proposal. Lots of waters have flowed over this proposal, and it is now clear that time has surpassed it. Indeed, when Egypt and Sudan turned to this offer, and were considering it in 2016, Ethiopia indicated that it was too late – the dam was close to 60 per cent complete.

Six years after commencement of construction of the GERD, the dam has no doubt become a reality, cemented by two instruments, one of them signed by the presidents of Egypt and Sudan and the prime minister of Ethiopia, themselves. Clearly, the GERD is poised to be completed according to the original schedule and time – at the end of 2017.

The GERD evidently reflects the reality of the changing power paradigm in the Nile Basin, as well as the emergence of a resulting new legal order. This order is manifested in the DoP, and confirmed by the Khartoum Document, replacing for all practical and legal purposes, the 1902 Treaty and the 1959 Nile Waters Agreement.

This conclusion is buttressed by a number of legal facts. First and foremost, the DoP does not reiterate Egypt's claim of the validity of the 1902 Treaty; it does not even mention the Treaty. Moreover, it does not refer to Egypt and Sudan's demand for a clear recognition of their water security. This demand was presented and insisted upon during negotiations of the Nile Basin Cooperative Framework Agreement (CFA), when Egypt and Sudan pressed hard for the inclusion of explicit provisions regarding their 'existing uses and rights'. Both countries have continued to oppose the CFA because of the absence of these provisions. Indeed, the DoP has underscored repeatedly the basic principle of international water law of equality of all the riparians in the sharing and uses of the international watercourse. The DoP has further incorporated and cemented this principle through the adoption of the concept of equitable and reasonable utilisation and its factors, and through subordination of the no-harm obligation to this principle. Another incorporation by the DoP of the principles of international water law is the repeated calls in the DoP for cooperation among the Nile riparians. The principles set forth in the DoP were unequivocally confirmed nine months later by the Khartoum Document. Clearly, the playing field of the Nile Basin has been levelled.

Yet, the levelling of the playing field should help the Nile Basin countries, both downstream and upstream, realise the basic fact that the Nile River can only be properly managed through the cooperation and collective action of all the riparians. Given the large and increasing number of people living and depending on the Nile, the growing competing demands of its eleven riparians over the Nile water resources, and the rising challenges of climate change and environmental degradation facing the river, the Nile Basin is indeed in urgent need of concrete plans and actions in the direction of cooperation and collective action.

As the UN Watercourses Convention, the global shared watercourses tenets, expressly stipulates, it is only through cooperation that development, conservation, management and protection of international watercourses, and the promotion of the optimal and sustainable utilisation thereof for present and future generations, can be assured.

Notes

1 It should be noted that Egypt was able to fill the AHD reservoir, which has a capacity of 162 BCM, without major negative impacts to its huge irrigated agriculture. The capacity of the GERD reservoir is 74 BCM, about 40 per cent of that of the AHD.

2 The Ethiopia–Sudan inter-connector (194 km transmission interconnection between Bahr Dar and Gondar in Ethiopia; and 321 km connecting Gondar-Shehdi-Metema in Ethiopia with Gadaref in Sudan) has generated a number of benefits for the Sudan. Fully commissioned at the end of 2013, the interconnector with a capacity of 100 MW has enabled nearly 1.4 million households (both in Sudan and Ethiopia) to access reliable electricity (Nile Basin Initiative 2016).

3 It is reported that: 'Egypt is set to benefit from the on-going Eastern Nile Regional Transmission Line; Ethiopia–Sudan (Rabak)–Egypt (Nage-Hamad). The Ethiopia–Egypt line in particular will provide 2,000 MW, or 7,700 MWh/year' (Nile Basin Initiative 2016).

4 'Apart from improved reliability of supply, consumers have gained from lower tariffs of US\$0.05 per kWh for imported electricity as compared to US\$0.096 per kWh from power generated domestically' (Nile Basin Initiative 2016).

5 Some reports refer to the Tripartite National Committee (TNC) as the Technical National Committee (TNC); apparently because both carry the same abbreviation.

6 The Saudi delegation visit to the GERD angered the Egyptians, and the Egyptian government accused Saudi Arabia of being willing to ally itself 'with anyone in case Egypt does not comply with Saudi foreign policy' (Daily News Egypt 2016). This is quite surprising given that a number of Egyptian delegations, including the Minister of Irrigation and Water Resources of Egypt, had already visited the GERD.

7 The International Court of Justice in the Gabcikovo–Nagymaros case quoted from the 1929 judgement by its predecessor, the Permanent Court of International Justice (PCIJ), where the PCIJ stated:

> [the] community of interest in a navigable river becomes the basis of a common legal right, the essential features of which are the perfect equality of all riparian States in the user of the whole course of the river and the exclusion of any preferential privilege of any one riparian State in relation to the others.

See case relating to the Territorial Jurisdiction of the International Commission of the River Oder, *United Kingdom, Czechoslovakia, Denmark, France, Germany and Sweden* v. *Poland*, Judgement No. 16 (PCIJ., Ser. A, No. 23, 1929)

8 The UN Watercourses Convention entered into force on 17 August 2014, ninety days after deposit by Vietnam of the 35th instrument of ratification/accession of the Convention. As of August 2017, thirty-six States are parties to the Convention (Salman 2015).

9 The Nile Basin Cooperative Framework Agreement (CFA) is the agreement that was negotiated as the main component of the Nile Basin Initiative (NBI) (CFA 2010). It was signed in 2010 by Kenya, Uganda, Tanzania, Ethiopia and Rwanda. Burundi joined in 2011. Thus far, Ethiopia, Rwanda and Tanzania have ratified the CFA. The CFA needs six instruments of ratification to enter into force. Egypt and Sudan oppose the CFA vehemently (Salman 2013b).

10 This article of the DoP has clearly been the result of the MIT report, discussed above, urging coordination between the GERD and AHD and the Sudanese dams.

11 In fact, the three ministers had previously visited the GERD dam site. As discussed before, the visit took place on 25 September 2014, following the end of the tripartite meeting of the ministers of water resources in Addis Ababa, Ethiopia on 23 and 24 September 2014.

12 One feddan equals 0.42 hectares, or 1.38 acres.

References

Abdalla, I.H. (2000). Removing the Nubians: The Halfawis at Khashm al-Girba. In H. Erlich and I. Gershoni (eds), *The Nile: Histories, Culture, Myths* (pp. 235–244). Boulder, CO: Lynne Reinner Publishers.

Al Arabiya (2013, 4 June). Caught on Camera: Egyptian Politicians talk Covert Ethiopian Attack. Retrieved from http://english.alarabiya.net/en/News/middle-east/2013/06/04/Egyptian-politicians-suggest-sabotaging-Ethiopia-s-new-Nile-dam.html.

Blackmore, D. and Whittington, D. (2008). Opportunities for Cooperative Water Resources Development on the Eastern Nile: Risks and Rewards. An Independent Report of the Scoping Study Team to the Eastern Nile Council of Ministers.

Cascão A. and Nicol, A. (2016). GERD: New Norms of Cooperation in the Nile Basin?. *Water International, 41*(4), 550–573.

CFA (2010). The Nile Basin Cooperative Framework Agreement (CFA). Retrieved from www.nilebasin.org/images/docs/CFA%20-%20English%20%20FrenchVersion.pdf.

Collins, R. (1996). *The Waters of the Nile: Hydropolitics and the Jonglei Canal, 1900–1988*. Princeton, NJ: Markus Wiener Publishers.

Collins, R. (2002). *The Nile*. New Haven, CT: Yale University Press.

Dafalla, H. (1975). *The Nubian Exodus*. London: Hurst and Uppsala: Scandinavian Institute of African Studies.

Daily News Egypt, (2016, 18 December). Saudi Official Reported Visit to GERD could Deepen Tensions with Egypt. Retrieved from www.dailynewsegypt.com/2016/12/18/saudi-official-reported-visit-gerd-deepen-tensions-egypt/.

Degefu, G.T. (2003). *The Nile: Historical, Legal and Developmental Perspectives*. Victoria, BC: Trafford Publishing.

DoP (2015). Agreement on Declaration of Principles between the Arab Republic of Egypt, the Federal Democratic Republic of Ethiopia, and the Republic of the Sudan on the Grand Ethiopian Renaissance Dam Project (GERDP). Retrieved from http://hornaffairs.com/en/2015/03/25/egypt-ethiopia-sudan-agreement-on-declaration-of-principles-full-text/.

Ezega (2017, 25 February). Ethiopian Renaissance Dam Generation Capacity Revised Up to 6,450MW. Retrieved from www.ezega.com/News/NewsDetails/3978/Ethiopian-Renaissance-Dam-Generation-Capacity-Revised-Up-to-6-450MW%5C.

Eleiba, A. (2013). Sudanese president backs Ethiopian dam ahead of Nile talks. *Ahram Online*, 5 December. Retrieved from http://english.ahram.org.eg/NewsContent/2/8/88321/World/Region/Sudanese-president-backs-Ethiopian-dam-ahead-of-Ni.aspx.

Garretson, A.H. (1967). The Nile Basin. In A.H. Garretson, R.D. Hayton and C.J. Olmstead (eds), *The Law of International Drainage Basins* (pp. 256–297). Dobbs Ferry, NY: Oceana Publications.

International Panel of Experts (2013). Grand Ethiopian Renaissance Dam, Final Report, Addis Ababa, Ethiopia, 31 May.

Khartoum Document (2015). Summary and the Outcomes of the Meeting, on File with Author.

Mekonnen, Y. (1984). *The Nyerere Doctrine of State Succession and the New States of East Africa*. Arusha, Tanzania: Eastern Africa Publications.

MIT (2014). The Grand Ethiopian Renaissance Dam: An Opportunity for Collaboration and Shared Benefits of the Eastern Nile Basin – An Amicus Brief to the Riparian Nations of Ethiopia, Sudan and Egypt from the International, Non-Partisan Eastern Nile Working Group Convened at the Massachusetts Institute of Technology on

13–14 November 2014 by the MIT Abdul Latif Jameel World Water and Food Security Lab.

Nile Basin Initiative (2016). Building on Shared Benefits: Transforming Lives in the Nile Basin. Entebbe, Uganda, at 36.

Okidi, O. (1994). History of the Nile and Lake Victoria Basins Through Treaties. In P.P. Howell and J.A. Allan (eds), *The Nile: Sharing a Scarce Resource* (pp. 321–350). Cambridge: Cambridge University Press.

Press Release (2016). Press Release of the 10th Tripartite National Committee: TNC Meeting on the Grand Ethiopian Renaissance Dam (GERD).

Salman, S.M.A. (2011). Grand Ethiopian Renaissance Dam: Challenges and Opportunities. *The CIP Report* (Center for Infrastructure Protection), 10(4), 21–29. University of George Mason, School of Law Retrieved from https://cip.gmu.edu/wp-content/uploads/2013/06/111_The-CIP-Report-October-2011_DamSector.pdf.

Salman, S.M.A. (2013a). The Ethiopian Renaissance Dam: Opportunities & Challenges, *Sudanow Magazine*, Retrieved from http://sudanow.info.sd/the-ethiopian-renaissance-dam-opportunities-challenges/.

Salman, S.M.A. (2013b). The Nile Basin Cooperative Framework Agreement: A Peacefully Unfolding African Spring?. *Water International*, 38(1), 17–29.

Salman, S.M.A. (2014). Sudan Continues Relinquishing a Growing Portion of Nile Waters Share. Retrieved from www.salmanmasalman.org/sudan-continues-relinquishing-a-growing-portion-of-nile-water-share/.

Salman, S.M.A. (2015). Entry into Force of the UN Watercourses Convention: Why Should it Matter?. *International Journal of Water Resources Development*, 31(1), 4–16.

Salman, S.M.A. (2016). *Sudan and the Nile Waters*. Fairfax, VA: Sudan Research Center (in Arabic).

Tawfik, R. (2016). The Grand Ethiopian Renaissance Dam: A Benefit-Sharing Project in the Eastern Nile. *Water International*, 41(4), 574–592.

United States Bureau of Reclamation (USBR) (1964). Land and Water Resources of the Blue Nile Basin: Ethiopia, Main Report and Appendices I–V. Washington, DC: United States Department of Interior; US Government Printing Office.

Waterbury, J. (2002). *The Nile Basin: National Determinants for Collective Action*. New Haven, CT: Yale University Press.

Wheeler, K.G., Basheer, M., Mekonnen, Z.T., Eltoum, S.O., Mersha, A., Abdo, G.M., Zagona, E.A., Hall, J.W. and Dadson, S.J. (2016). Cooperative Filling Approaches for the Grand Ethiopian Renaissance Dam. *Water International*, 41(4), 611–634.

World Bank (2006). Ethiopia: Managing Water Resources to Maximize Sustainable Growth: A World Bank Water Resources Assistance Strategy for Ethiopia. Washington, DC.

Al-Youm, A. (2017, 9 April). Qatari Emir to visit Ethiopia on Monday. *Egypt Independent*. Retrieved from www.egyptindependent.com/news/qatari-emir-visit-ethiopia-Monday.

Zhang Y., Erkyihun, S.T. and Block, P. (2016). Filling the GERD: Evaluating Hydroclimatic Variability and Impoundment Strategies for Blue Nile Riparian Countries. *Water International*, 41(4), 593–610.

4 International law developments on the sharing of Blue Nile waters

A fairness perspective

Zeray Yihdego and Alistair Rieu-Clarke

Introduction

The principle of fairness operates alongside lofty principles of international law, such as equity and justice. However, these concepts often face criticism for being too vague to shed any meaningful light on the practical interpretation and implementation of international law within specific fields. By analysing the cooperation between Egypt, Ethiopia and Sudan on the Blue Nile, this paper seeks to examine such criticism. The chapter suggests that the concept of fairness does have value as a framework for analysing both commitment and compliance in international law; and that exploring specific contexts, such as legal developments related to the Grand Ethiopian Renaissance Dam and relevant (legal) instruments, helps give it an objective and normative meaning. The chapter will also show how the realization and compliance with principles of (international) law such as the fairness principle require an input from other disciples – in this chapter's case, the input from economics and hydrology have been used to try to objectively determine distributive justice as one crucial element of fairness.

Fairness, a well-established concept in law, is often closely associated with the principles of equity and justice. Rawls maintains that, 'the fundamental idea in the concept of justice is fairness' (1958: 164); Garner defines justice as, 'the fair and proper administration of the law' (1999: 881), while describing equity as, 'fairness; impartiality; evenhanded dealing' (1999: 579). Chapman draws the distinction between 'fairness being concerned with process, e.g. a fair trial, and justice relating to the outcome, e.g. a just decision' (Chapman 1963: 154). Judge Owada suggests that: 'considerations of fairness in the administration of justice requires equitable treatment of the positions of both sides involved in the subject-matter in terms of the assessment both of facts and of law involved' (Legal Consequences of the Construction of a Wall in the Occupied Palestinian Territory 2004: 260).

As these opinions demonstrate that the distinction between fairness, equity and justice is not always clear. Despite the difficulties in defining fairness and demarcating its meaning from equity and justice, it has proven to be a popular tool by which to analyse various legal fields, including climate change (Soltau 2009), environmental protection (Louka 2006; Franck 1995: 380–412; cf. Hsu

2004), sustainable development (Brown-Weiss 1990), trade and investment (Franck 1995: 413–473), international arbitration (Desierto 2015; Sarvarian *et al.* 2015) and corporate law (Mitchell 1993). Perhaps one of the most attractive facets of the concept of fairness is its perceived correlation to compliance. Franck, for instance, maintains that the degree to which a particular rule or principle is perceived to be fair will influence its 'compliance pull' (Franck 1988: 706). A study of fairness, therefore, transcends an examination of what is, or what is not, international law, and rather wrestles with the arguably more pertinent question of why international legal obligations are obeyed.

Given the rich literature on fairness, both in general theory and within specific legal fields, it is somewhat surprising that the concept has received little attention within the legal field of international watercourses. This paper seeks to rectify this shortcoming by suggesting where considerations of fairness may add value to the study the law of international watercourses; and also, where insights from the law of international watercourses, and cooperation over the Nile River Basin in particular, may inform more general ideas of fairness.

The paper is divided into three main sections. The first section outlines the linkages between the principle of fairness and the law of international watercourses. While numerous interpretations of the concept of fairness have been offered by scholars, this chapter concentrates on the seminal work of Thomas Franck. The second section analyses the specific case of the Nile River Basin. Through a 'fairness lens', the section explores how the concept of fairness can help explain the strengths and weaknesses of the legal regime relating to the Nile. The section focuses primarily on the key legal instruments that have shaped, or reflected, cooperation on the Nile in general, and the Eastern Nile Basin in particular. The intention is to tease out examples of where considerations of fairness may have relevance to the creation, interpretation and application of international law relating to the Nile River Basin. The final section draws upon the two previous sections in order to offer insights for policy makers and academics with an interest in the development of the Nile legal framework, and those more generally interested in the relationship between fairness and international law.

Applying fairness to the law of international watercourses

Building on Franck's concept of fairness (1995), which he examines within the context of international law and institutions, offers a framework by which to study fairness within the context of the law of international watercourses.

Franck suggests that,

> the fairness of international law, as of any other legal system, will be judged, first by the degree to which the rules satisfy the participants' expectations of justifiable distribution of costs and benefits, and secondly by the extent to which the rules are made and applied in accordance with what the participants perceive as right process.
>
> (Franck 1995: 7)

In offering this interpretation, Franck envisages that fairness is comprised two key elements: legitimacy and distributive justice (Franck 1995: 7–9).

Franck recognises the importance of legitimacy in the effective implementation of legal rules by suggesting that, 'in a community organised around rules, compliance is secured – to whatever degree it *is* – at least in part by a perception of a rule as legitimate by those to whom it is addressed' (Franck 1988: 706). What constitutes a legitimate rule, according to Franck, is one that '*derives from a perception on the part of those to whom it is addressed that it has come into being in accordance with right process*' (emphasis in the original) (Franck 1988: 706). Legitimacy, according to Franck, can also be described as 'procedural fairness' (Franck 1995: 7), which demands that proper processes are in place to create, interpret and apply international law (Scobbie 2002: 910). This begs the question what might be deemed 'right process', or 'proper processes'? Franck provides some suggestions, by stating that: 'decisions about distributive and other entitlements will be made by those duly authorised, in accordance with procedures which protect against corrupt, arbitrary, or idiosyncratic decision-making or decision-executing' (1995: 7). While Franck therefore provides some pointers, what might constitute 'right process' within any given context requires further consideration.

The law relating to international watercourses recognises 'right process' in various procedural obligations concerning the creation, interpretation and application of international law (McIntyre 2010; Ziganshina 2015). These procedural requirements are founded upon a general requirement that, '[w]atercourse States shall cooperate on the basis of sovereignty equality, territorial integrity, mutual benefit and good faith in order to attain optimal utilisation and adequate protection of an international watercourse' (UNWC 1997; see also Leb 2013). In adding texture to this general duty to cooperate, the *Convention on the Law of the Non-Navigational Uses of International Watercourses* ('UN Watercourses Convention' or 'UNWC'), goes on to set out detailed procedural requirements by which states must regularly exchange data and information (UNWC 1997: Art. 9), notify and consult on any planned measures, including the sharing of any environmental impact assessment (UNWC 1997: Arts 11–19), and settle disputes in a peaceful manner (UNWC 1997: Art. 33). The UN Watercourses Convention also infers the need for process when it requires watercourse states to, '*participate* in the use, development and protection of an international watercourse in an equitable and reasonable manner' (emphasis added) (UNWC 1997, Art. 5[2]); 'take *appropriate measures* to prevent the causing of significant harm to other watercourse States' (emphasis added) (Art. 7[1]); cooperate on the prevention, reduction and control of pollution of an international watercourse (Art. 21[3]); prevent the introduction of alien or new species (Art. 22); protect and preserve the marine environment (Art. 23), manage and regulate an international watercourse (Arts 24 and 25); maintain and protect installations (Art. 26); prevent and mitigate harmful conditions (Art. 27); and manage emergency situations (Art. 28). However, while the importance of process is implicit within these provisions, the detail is largely lacking. This, therefore, begs the question, whether a stronger focus and clarity on procedural fairness would help strengthen the implementation of the UN Watercourses Convention. The analysis of the Nile River Basin below will seek to explore where such clarity might be provided.

A further area where greater clarity might be gained relates to the law-making process. The requirement to cooperate in *good faith* under Article 8(1) of the UN Watercourses Convention, suggests that there are certain obligations upon states during the negotiation of legal arrangements (Leb 2013: 31–32). This requirement is supplemented by, Articles 3 and 4 of the UN Watercourses Convention, which requires that states to negotiate watercourse agreement in good faith, or adjustments thereof (UNWC 1997). However, despite good faith being an important benchmark to apply during negotiations, it is not defined in the UN Watercourses Convention. Recourse must therefore be made to general guidance, such as that offered by the International Court of Justice (ICJ) when it states that good faith negotiations between parties must be, 'meaningful, which will not be the case when either of them insists upon its own position without contemplating modifications to it' (North Sea Continental Shelf Cases 1969: 47; see also O'Connor 1991; D'Amato 1992; Liguori 2009). Similarly, the Arbitral Tribunal in the *Lake Lanoux Arbitration* suggested that the obligation to negotiate in good faith would be breached in cases, 'of an unjustified breaking off of the discussions, abnormal delay, disregard of the agreed procedures, systematic refusals to take into consideration adverse proposals or interests' (Lake Lanoux Arbitration 1957: 23). While this general guidance is no doubt useful in guiding state negotiations, more could be done to expand on these general explanations of good faith, and provide specific guidance for states and other actors involved in the negotiation of treaty arrangements relating to international watercourses.

The second element of fairness, as set out by Franck, is distributive justice. Franck (1995) associates distributive justice with substantive issues concerning the allocation of resources. He suggests that:

> when everyone can expect to have a share, but no one can expect to have all that is desired, the critical moment for considerations of fairness is met. It is only then that modes of allocation do not contend in the arena of a zero-sum game, one that pits the survival of each against the survival of all.
>
> (Franck 1995: 10)

Franck's observation shares familiar ground with the International Law Commission's (ILC) explanation of how equitable utilisation should be applied. The ILC commentary to the 1994 ILC *Draft Articles on the Law of the Non-navigational Uses of International Watercourses* (1994 Draft Articles), suggests that,

> where the quantity or quality of the water is such that all the reasonable and beneficial uses of all watercourse States cannot be fully realised, a

'conflict of uses' results. In such a case, international practice recognises that some adjustments and accommodations are required in order to preserve each watercourse State's equality of right. These adjustments or accommodations are to be arrived at on the basis of equity, and can best be achieved on the basis of specific watercourse agreement.

(ILC 1994: 98)

From this explanation, it would appear that Franck's notion of distributive justice aligns well with the principle of equity in international law. Like the ILC in its commentary to the 1994 Draft Articles, Franck does not use equity primarily in its broadest sense of filling gaps in the law, or as a general principle of law (Akehurst 1976: 801–802; McIntyre 2013: 113–117). A broad interpretation of equity may face criticism. Jennings for instance, describes applying equity to fill gaps in the law as tantamount to 'subjective appreciation' of facts and laws in dealing with legal claims and counter-claims (Jennings 1986: 31). Different from this, Franck sees distributive justice, like equitable utilisation, as a substantive rule of apportionment (McIntyre 2013: 120–124), which arguably has more traction. Franck (1995), suggests that such a substantive rule rejects making absolute or non-negotiable claims, which as noted earlier in the comments by the ICJ, would be at variance with the principle of good faith.

The principle of equitable utilisation, as reflected in the UN Watercourses Convention (UNWC 1997: Art. 5) is a well-established principle of customary international law. In addition, there has been a considerable amount of scholarly work dedicated to examining equity within the context of international watercourses and, more generally to the allocation of shared resources (e.g. Fuentes 1996; Kaya 2003; Lipper 1967; McIntyre 2013; McCaffrey 2007: 384–405). What, then, is the benefit of looking at equity through a fairness lens? The notion of distributive justice, as set out by Franck, adds three important aspects.

First, by placing equitable utilisation within a fairness framework, Franck recognises the importance of distributive justice to compliance. Franck maintains that: 'the perception that a rule or system is distributively fair, like the perception of its legitimacy, … encourages voluntary compliance' (Franck 1995: 8).

Second, in Franck's articulation of distributive justice, there is an implication that allocation must not only be between states, but among a community, 'based … on a common, conscious system of reciprocity between its constituents' (Franck 1995: 10). While Franck's own application of legitimacy has been criticised as being overly positivist and focused on states (Brunnée and Toope 2002: 52–53), a study of international watercourses provides an appropriate space by which to move beyond states and consider all constituents within a particular basin. Taking this further, it could be argued that distributive justice must account for multiple scales in allocating the uses and benefits that derive from international watercourses (Patrick 2014). However, the law of international watercourses has traditionally been skewed towards state–state relations. Article 5 of the UN Watercourses Convention on equitable utilisation, for instance, provides that, 'Watercourse *States* shall in their respective territories utilise an international watercourse in an equitable and reasonable manner' (emphasis added) (UNWC 1997).

In relation to equitable utilisation, the only reference to non-state actors is framed in terms of factors that states must account for in their determination of what is equitable and reasonable, such as the population dependent on a water-course and their social and economic needs (Art. 6) or vital human needs (Art. 10). As a counter to this state–state bias, human rights law is playing an increasingly influential role in relation to issues of access to water and adequate sanitation provision (UN GA Res. 2010; UN CESCR 2002; Salman 2004). To date, much of the focus of the right to water has been at a domestic level. More, therefore, needs to be done to understand the implications of the human right to water within the world's transboundary rivers, lakes and aquifers – especially given that 40 per cent of the world's population rely on these shared waters (Archer 2012; Bulto 2013; Leb 2012; McCaffrey 1992). A community-centric approach to distributive justice, which encompasses both state and non-state actors, offers potential to draw together these traditional disparate areas of the law of international watercourses. It ought to be noted that this line of argument is compatible with Rawls's law of peoples (1971), which expanded the concept of justice as fairness to an international arena.

Third, the importance of recognising the interdependency between legitimacy and distributive justice is central to Franck's concept of fairness. In this regard, Frank comments that,

> [l]egitimacy and distributive justice are two aspects of the concept of fairness. While one has a primary procedural, and the other a primarily moral perspective, they combine to answer the law-maker's version of the question posed by Socrates and Jeremy Bentham: 'What shall we do about sharing and conserving in order to maximise human well-being?'
>
> (Franck 1995: 8–9)

However, Franck cautions that the two elements might not necessarily pull in the same direction given that distributive justice 'favours change', and legitimacy favours 'stability and order' (Franck 1995: 7). Franck suggests that, '[t]he tension between stability and change, if not managed, can disorder the system. Fairness is the rubric under which this tension is discursively managed' (Franck 1995: 7).

The correlation between distributive justice and legitimacy resonates well with the law relating to international watercourses. This important correlation was recognised by the International Court of Justice in the *Case Concerning Pulp Mills on the River* Uruguay through the discussion of procedural norms (including notification, environmental impact assessment and consultation), and the substantive requirements of a bilateral treaty between Argentina and Uruguay (2010: 37–39). However, the need to link distributive justice to legitimacy is often missing in the design of treaties relating to international watercourses (see for example, Garane and Abdul-Kareem 2013; Ziganshina 2013), even though the determination of equitable utilisation is heavily reliant on the 'right processes' being in place to ascertain and reconcile competing needs and interests within a particular watercourse.

Through the introduction of the concept of equitable participation within the UN Watercourses Convention, the importance of linking substantive norms to process was not lost on the ILC (ILC 1994: 97). In their commentary to the 1994 Draft Articles, the ILC maintained that, the affirmative nature of the principle of equity encompassed not only, 'the right to utilise the watercourse', but also the duty to cooperate actively with other watercourse states 'in the protection and development' of the watercourse (ILC 1994: 97). This articulation of equitable *participation*, which found expression in Article 6 of the UN Watercourses Convention, demonstrates a clear attempt by the ILC to link the substantive principle of equitable utilisation, with the procedural requirements inherent in the duty to cooperate (as discussed above).

The linkages between substantive norms and process are also evident within the duty to take all appropriate measures to prevent significant harm (UNWC 1997: Art. 7). This so-called 'due diligence' obligation, asks whether 'appropriate measures' were put in place to prevent that harm (Wouters and Chen 2013: 236). Such measures might include procedural requirements to notify and consult (UNWC 1997: Arts 11–19), as well as those related to pollution, for example, joint water quality objectives and criteria, techniques and practices to address pollution from point and non-point sources, and the creation of lists of substances to be banned, limited, investigated and/or monitored (UNWC 1997: Art. 21[3]). The ICJ has also suggested that:

> due diligence ... would not be considered to have been exercised, if a party planning works liable to affect the régime of the river or the quality of its waters did not undertake an environmental impact assessment on the potential effects of such works.
>
> (Case Concerning Pulp Mills on the River Uruguay 2010: 83)

Further guidance can be sought from the Convention on Protection and Use of Transboundary Watercourses and International Lakes (UNECE WC 1992), which sets out detailed legal, financial and technical measures that states should take to prevent, control or reduce 'transboundary impact'.

This section has demonstrated that an analysis of fairness within the context of international watercourses has its merits. The value in combining an analysis of process, in broad terms, with issues of substance, is a clear feature in the study of fairness, and critical to an understanding of why laws generate 'compliance pull' (Franck 1988: 706). Through the above analysis of the law of international watercourses, the importance of linking substantive considerations of distributive justice to procedural considerations is evident. However, it has also been shown that there are some notable areas where the linkages could be strengthened. Three key areas are noteworthy. First, laws relating to international watercourses, as shown above, are rule-based, whereas Franck regards distributive justice and legitimacy as a moral and legal concepts, respectively. This in turn means that a study of the law of international watercourses through a fairness lens can focus greater attention on the factors that influence compliance. Second, the law of international watercourses is still primarily concerned with state–state relations and therefore fails to pick up on the broader community-oriented and multi-level aspects of distributive justice. Third, while the importance of 'right process' with respect to law creation are captured in general principles such as good faith, more detail is required in order to guide states and other actors when they negotiate water-specific treaty arrangements. These key areas will be considered further in the following section, by exploring fairness in relation to legal developments on the Blue Nile.

Fairness and the Blue Nile

Early Nile treaties and fairness

Several treaties relating to the Nile waters were concluded from the end of the nineteenth century to the 1950s. One of the most significant of those treaties, in terms of fairness considerations, was the 1902 Treaty between Great Britain (on behalf of Sudan) and Ethiopia (Nile Treaty 1902). While the treaty primarily focused on establishing the border between Ethiopia and Sudan, it also required Ethiopia,

> not to construct or allow to be constructed, any work across the Blue Nile, Lake Tsana, or the Sobat, which would arrest the flow of their waters except in agreement with His Britannic Majesty's Government and the Government of Soudan.
>
> (Nile Treaty 1902: Art. III)

Article III of the 1902 Nile Treaty is among the key colonial treaty provisions which were, and still are, invoked to oppose the construction of any project on major Nile tributaries stemming from Ethiopia (Shetewy 2013: 34–35).

Then, in 1929, an agreement was concluded between the United Kingdom (again on behalf of Sudan) and Egypt relating to the, 'Use of the Waters of the River Nile for Irrigation Purposes' (Nile Treaty 1929). The text of the 1929 Nile Treaty stated that:

> Except with the previous agreement of the Egyptian Government, no irrigation of power works or measures are to be constructed or taken on the River Nile and its branches, or on the lakes for which it flows, so far as all these are in the Sudan or in countries under British administration, which would, in such a manner as to entail any prejudice to the interests of Egypt, either reduce the quantity of water arriving in Egypt, or modify the date of its arrival, or lower its level.
>
> (Nile Treaty 1929: Art. 4[ii])

Thirty years later, and despite Ethiopia serving notice in 1957 that it would uni-laterally develop the Nile water resources within its territory, Egypt entered into a treaty with by then independent Sudan that was boldly titled 'The Full Utili-sation of the Nile Waters' (Nile Treaty 1959). The 1959 treaty allocated the entire available average annual flow of the Nile between Sudan (18.5 billion cubic metres) and Egypt (55.5 billion cubic metres) (see McCaffrey 1992: 269; Degefu 2003: 99).

The conclusions of these bilateral arrangements raise a number of issues relating to legitimacy and distributive justice. A key question concerns whether the states can rely upon, or be obliged to comply with, a treaty that was entered into by a former colonial power. While state succession to treaties is a contested area of international law, the *Vienna Convention on the Succession of States in Respect of Treaties* stipulates that:

> [a] newly independent State is not bound to maintain in force, or to become a party to, any treaty by reason only of the fact that at the date of the suc-cession of States the treaty was in force in respect of the territory to which the succession of States relates.
>
> (Vienna Convention 1978: Art. 16)

This provision supports a wider doctrine of 'clean state' or 'non-devolution' that suggests that all (bilateral) treaties, especially those which lack universal appeal, are not assumed to be succeeded by independent or newly created states (Maluwa 1999: 69; Leb and Tignino 2013: 425).

One suggested exception to this 'clean state' rule relates to treaties that establish boundaries (Shaw 1996: 97, 104). Article III of the 1902 Treaty was part of a boundary treaty. However, if one looks into the history of the provi-sion, it is clear that, in order to protect Britain's interests in the Nile, it was sep-arately negotiated for about three years before being inserted into the 1902 Treaty at the last minute (Woldestadik 2013: 100; Garretson 1967: 277). Addi-tionally, Article III restricts activities concerning an international watercourse, rather than stipulating that particular piece of land, lake or territorial waters within a state's sovereign territory (Shaw 1996: 77). It would, therefore, seem to fall outside the boundaries exception.

While the 'clean state' doctrine may therefore be relied upon to question the legitimacy of the 1902 and 1929 Treaties with respect to Sudan, Ethiopia was an independent state when it entered into the 1902 Treaty (Aal 2013: 27). This might preclude Ethiopia setting forth arguments based on succession, although other issues relating to fairness have been raised. The unevenness of the deal between a colonial power and a country under exigent threat from different colonial powers may raise questions of 'right process' (Degefu, 2003: 99). The Nile colonial-era treaties, including the 1902 Treaty, were instigated by the self-interest of colonial powers rather than the shared interest of the community of Nile riparian states. For this reason, and despite the claim of pursuing 'a civilis-ing mission', such colonial treaties, have been 'discontinued in international law' as substantively flawed and discriminatory (Boyle and Chinkin 2007: 28–29). Moreover, the circumstances of concluding the 1902 Treaty, including its ambiguity, subsequent rejection and non-ratification by Ethiopia are significant factors to take into account when questioning its procedural fairness.

The early Nile Treaties might also be questioned from the standpoint of distributive justice. As noted in the previous section, distributive justice is founded upon the notion of equity, which, in the context of international watercourses, requires waters to be utilised in an equitable and reasonable manner (UNWC 1997: Art. 5). Both the 1929 and 1959 Agreements may be questioned on the basis of distributive justice given that they effectively aim to allocate the waters of the Nile between the two most downstream states, without seeking to account for the 'community' of relevant factors and circumstances, including the needs and interests of upstream states. The exclusion of upstream states during their negotiation might also raise a further question of procedural fairness; and would certainly be at variance with Article 4 of the UN Watercourses Convention and the entitlement of watercourse states to participate in the negotiation of an agreement that applies to the entire international watercourse (UNWC 1997).

Assessing the fairness of such old Nile treaties in light of the UN Watercourses Convention may be questioned on the basis that none of Nile Basin states are parties to that Convention. However, there appears consensus that the basic substantive and procedural duties contained in the Watercourses Convention reflect customary international law in the field and are thus applicable to all states (McCaffrey 2007). Additionally, the legitimacy of those provisions and their 'compliance pull' may be secured by a more explicit recognition of key rules and principles by Nile Basin states, and arguably other actors in the basin. One explicit recognition would be if the Nile Basin states became party to the UN Watercourse Convention. Short of that, it could be argued, as will be discussed below, that some of the key provisions of the UN Watercourses Convention have been endorsed within cooperative instruments relating to the basin. Finally, the 1902 Nile Treaty also illustrates the correlation between distributive justice (substantive) and legitimacy (procedure). While the right to veto plans in Article III of the Treaty is a procedural requirement, it is difficult not to think that it might hinder achieving an equitable solution. Such a right runs the risk of jeopardising an exercise wherein all relevant factors and circumstances are taken into account, and weighed on the basis of the whole, in the determination of what might constitute an equitable use of an international watercourse.

Post-1959 developments in the Nile and fairness

While the earlier Nile treaties proved controversial among the Nile Basin states, attempts were made to find a more cooperative path in the later part of the twentieth century (Brunnée and Toope 2010: 108–110). That effort resulted in the adoption of the *Framework for General Co-operation between the Arab Republic of Egypt and Ethiopia* (1993). The 1993 Framework recognised the 'mutual interest' in the Nile (preamble), together with their commitment to the principle of international law, including 'good neighbourliness', 'peaceful settlement of disputes', and 'non-interference in the internal affairs of states' (Framework 1993: Preamble and Art. 1). In terms of distributive justice, the 1993 Framework is short on detail. There is no reference to equity. More generally, Article 4 obliges the states to cooperate, 'on the basis of rules and principle of international law' (Framework 1993), and Article 5, stipulates that, 'each party shall refrain from engaging in any activity related to the Nile Waters that may cause appreciable harm to the interests of the other Party' (Framework 1993). Such an approach departs from the UN Watercourses Convention, which recognises that some harm may be tolerated if deemed equitable (UNWC 1997: Art. 7[2]). A question can therefore be raised over the interpretation of the 1993 Framework Instrument, and the extent to which distributive justice finds expression therein. Subsequent legal developments, which will be discussed below, would suggest a need to interpret the 1993 Framework in light of the principle of equitable utilisation.

The legitimacy of the 1993 Framework is also cause for debate (Abdo 2004: 51; Shetewy 2013: 35). Under Article 2(1)(a) of the Vienna Convention on the Law of the Treaties, a 'treaty', is described as 'an international agreement concluded between States in written form and governed by international law ...' (Vienna Convention 1969). Pursuant to this definition, the binding nature of agreements depends on whether it can be ascertained that the parties intended to be legally bound by the instrument (Crawford 2012: 372–373). In relation to its content, the 1993 Framework offers a mixed set of messages. While some provisions are couched in soft terms such as, 'the parties reaffirm their commitment' (Art. 1), 'the parties are committed to' (Art. 2), 'the parties recognise' (Art. 3), and 'the parties shall endeavour' (Art. 8); other provisions suggest stronger commitment, for example, 'the two parties agree' (Art. 4), 'the two parties shall refrain' (Art. 5), and 'the parties will create appropriate mechanisms' (Art. 7). At least some of the provisions of the Framework may, therefore, be said to be normative in character.

Examining the particular circumstances surrounding the Framework confounds the instruments uncertain status. Egypt registered the 1993 Framework within the UN. Such a practice is consistent with the requirement under Article 102(1) of the Charter of the United Nations (1945), which provides that, 'every treaty and every international agreement entered into by any Member of the United Nations ... shall *as soon as possible* be registered with the Secretariat and published by it' (emphasis added). Egypt's action would appear to demonstrate that it believed that the 1993 Framework was a legally binding treaty. The fact that Egypt relied upon the Framework, as well as the 1902 Nile Treaty, in formulating its position regarding the Grand Ethiopian Renaissance Dam (GERD) reinforces this position (Egyptian Ministry of Foreign Affairs n.d.). Conversely, the Treaty was not registered until seventeen years after its signing, which would hardly comply with the 'as soon as possible' requirement under the UN Charter. The legal significance of registration is also unclear. Unlike its predecessor, the League of Nations Covenant (Art. 18), non-registration of an agreement does not in itself decide whether an agreement is legally binding or not (Charter of the United Nations, 1945, Art 102[2]). The ICJ and other UN organs do not therefore consider registration of an instrument when dealing with cases involving agreements concluded between states (Martens 2012: 2109). Moreover, the framework has no procedures for its ratification, which, while not conclusive, offers an important sign of the legally binding nature of an agreement.

Regardless of whether or not the 1993 Framework is legally binding, the ambiguity surrounding its content and legal status raises serious questions over its legitimacy. Such questions are no doubt responsible for a lack of detailed follow-up by the states in implementing and ensuring their compliance with the requirements of the instrument. This record would therefore support Franck's argument that where questions of distributive justice and legitimacy are raised, the compliance pull of an instrument is likely to be low.

Following the 1993 Framework, the next major legal development related to the Nile was the establishment of the Nile Basin Initiative (NBI) in 1999 and the adoption of the Nile Cooperative Framework Agreement (CFA) (CFA 2010; Salman 2013: 19–20). The NBI was established by all Nile riparian states – except Eritrea, which had observer status – and was supported by the World Bank, the UN Development Programme and other donors. South Sudan joined the NBI on 5 July 2012. This intergovernmental organisation set out a shared vision, 'to achieve sustainable socio-economic development through equitable utilisation of, and benefit from, the common Nile Basin water resources' (NBI 2016b). Equity was therefore front and centre in the aspirations of the NBI. Cooperation between states through the NBI has led to the development of joint investment projects at sub-basin levels, coordinated by the Nile Equatorial Lakes Subsidiary Action Program Coordination Unit (NELSAP-CU), and the Eastern Nile Technical Regional Office (ENTRO) (Brunnée and Toope 2002). The NBI was designed to be a transitional arrangement towards a more comprehensive basin-wide legal and institutional framework (Salman 2013: 19). Adoption of the CFA was considered to be a critical milestone in that transition (Salman 2013: 19).

The text of the CFA was developed over more than a decade. The process began in 1997 with a panel of experts helping prepare text and background documents (NBI 2016a). Between 2000 and 2001, the text was converted into a draft Agreement by a transitional committee (NBI 2016a). Basin states then formed a negotiations committee from 2003 to 2005 in order to negotiate a draft agreement (NBI 2016a). This led to a draft text, which was submitted to the Council of Ministers of Water Affairs of the Nile Basin States (Nile-COM) in March 2006 (NBI 2016a). By June 2007, Nile-COM had completed its negotiations of the CFA, with all but one reservation – Article 14b related to Water Security (NBI 2016a). Attempts were made at various governmental levels, including with heads of state, to address the reservation, but to no avail

(Mekonnen 2010: 428). On 13 April 2000, seven countries agreed to open the CFA for signature. That decision was opposed by Egypt and Sudan (NBI 2016a). On 14 May 2010, four countries signed the CFA, with Kenya (19 May 2010) and Burundi (28 February 2011) following (NBI 2016a). To date Ethiopia (13 June 2013), Rwanda (28 August 2013), and Tanzania (26 March 2015) have ratified the CFA (NBI 2016a).

In terms of content, the CFA reflects a healthy balance between substantive norms and process. Through its incorporation of the principle of equitable and reasonable utilisation (Art. 4), the duty to take all appropriate measures to prevent significant harm (Art. 5), and the protection and conservation of the Nile River Basin and its ecosystems (Art. 6), the substantive duties of the CFA closely mirror the UN Watercourses Convention (CFA 2010). The balance between equitable and reasonable utilisation and the duty to take all appropriate measures to prevent significant harm, found in the UN Watercourses Convention, is also found in the CFA. One slight variation between the CFA and the UN Watercourses Convention concerns the factors and circumstances that should be taken into account when determining what is equitable and reasonable. Article 4 of the CFA offers two additional factors that should be considered, namely '[t]he contribution of each Basin State to the waters of the Nile River system', and '[t]he extent and proportion of the drainage area in the territory of each Basin State' (CFA 2010). However, the difference in the articulation of the relevant factors and circumstances in the two instruments is negligible given that both lists do not profess to be exhaustive.

Again, drawing on the UN Watercourses Convention, the CFA emphasises the importance of procedure for the implementation of substantive norms. Key procedural requirements contained in the CFA include the regular exchange of data and information (Art. 7), including in relation to planned measures (Art. 8), environmental impact assessment and audits (Art. 9), the establishment of the Nile Basin Commission (Arts 15–33), and the peaceful settlement of disputes (Art. 33) (CFA 2010). Notable from the standpoint of procedural fairness, and a requirement that is not explicitly contained in the UN Watercourses Convention, is the commitment that Nile Basin States shall, 'allow all those within a State who will or may be affected by the project in that State to participate in an appropriate way in the planning and implementation process' (CFA 2010: Art. 10). While it could be argued that this is an explicit statement of what is implied in the due diligence requirement to *take all appropriate measures* to prevent significant harm (CFA 2010: Art. 5; and UNWC 1997: Art. 7), its inclusion strengthens the importance assigned to the role of non-state actors within the implementation of cooperative arrangements.

Most of the text of the CFA might therefore be seen as consistent with the concept of fairness, at least in as much as the UN Watercourses Convention is. However, broader process-related issues and disagreements over Article 14 on Water Security raise questions over the legitimacy of the instrument. A major bone of contention throughout the negotiation of the CFA relates to the long-standing difference of opinion between upstream and downstream states on the

Nile concerning existing uses of and rights to the Nile Waters (Mekonnen 2010; Salman 2013: 21–22). Egypt and Sudan proposed that Article 14(b) should read: 'Nile Basin States … agree, in a spirit of cooperation … not to adversely affect the water security *and current uses and rights* of any other Nile Basin State' (emphasis added) (CFA: Article 14b and Annexe). Burundi, the Democratic Republic of Congo, Ethiopia, Kenya, Rwanda, Tanzania and Uganda proposed slightly different wording that omitted the reference to current uses and rights (CFA: Article 14b and Annexe).

From a fairness standpoint, the reference to current uses in the CFA might be seen as being at variance with distributive justice and the notion of equity. Franck recognises the importance of flexibility within the application of distributive justice when he cautions against the use of, 'a simple rule in circumstances requiring a more calibrated response' (Franck 1990: 77). Such a need for flexibility, in order to take account of the range of interests within a particular watercourse, as well as changes over time, is reflected in Article 10 of the UN Watercourses Convention, which reads: 'no use of an international watercourse enjoys inherent priority over other uses' (UNWC 1997). However, the trade-off between flexibility and 'compliance pull' is not lost on Franck. He stresses the importance of having effective mechanisms, or processes, in place so that any, 'ambiguity can be resolved case by case' (Franck 1988: 724). A major challenge for Nile Basin may therefore be to convince all states in the legitimacy of transboundary laws and institutions that can secure equitable outcomes in perpetuity without laying down rigid allocations of the waters of the Nile.

A major initial step in meeting this challenge will be to address the legitimacy issues surrounding the CFA. Egypt and Sudan did not approve the final text, and have not subsequently signed nor ratified the instrument (NBI 2016a). This is a significant bar on the emergence of basin-wide cooperation (McKenzie 2012: 571), which has led some to consider the CFA as untenable and also sparked a call to negotiate a new legal regime for the Nile (Kimenyi and Mbaku 2015: 83–89); others are calling for Sudan and Egypt to join in by asking the question: if the two endorsed the 2015 Declaration of Principles on the Ethiopian Renaissance Dam, why not the CFA? (Salman, Chapter 2). Irrespective of how this discussion will unfold, the process of adopting the CFA clearly demonstrates the importance of focusing on 'right process' in the negotiation and adoption of legal instruments. The adoption of the CFA without the consent of Egypt and Sudan can be seen as a major barrier to promoting the legitimacy of the CFA.

The Grand Ethiopian Renaissance Dam (GERD) and Fairness

While debates over the legal relevance and the fairness of the CFA continue, 'facts on the ground', in the shape of the GERD have also shaped cooperative arrangements within the basin. The GERD is under construction on the Blue Nile, approximately 20 kilometres from the Sudanese border in Ethiopia. Once complete, the project will be 145 m high, with an installed capacity of more than 6,000 megawatts. The energy produced by the project is intended for domestic consumption and export to other Nile riparian countries (Salman, Chapter 3). While Ethiopia considers the dam to be a key for its national development efforts, Egypt is concerned with its potential impact on its economy and its existing uses of the waters of the Nile (Tawfik and Dombrowsky, Chapter 5).

For this and other geopolitical reasons (Cascão and Nicol, Chapter 5), the dam was a source of tension between Egypt and Ethiopia before President El-Sisi came to power in June 2014. After then, painstaking negotiations, including the establishment of an International Panel of Experts (made up of six members from each country and four external experts), resulted in the adoption of the 'Declaration of Principles between The Arab Republic of Egypt, the Federal Democratic Republic of Ethiopia and the Republic of the Sudan on the Grand Ethiopian Renaissance Dam Project' ('DoPs') during March 2015 (DoPs 2015).

The DoPs is founded upon 'principles of cooperation', which include 'common understanding', mutual benefit', 'good faith', and 'principles of international law' (DoPs 2015: Art. I). Relevant principles of international law referred to in the text, include the 'Principle of Sovereignty and Territorial Integrity', whereby the three states commit to, 'cooperate on the basis of sovereign equality, territorial integrity, mutual benefit and good faith in order to attain optimal utilisation and adequate protection of the River' (DoPs 2015).

Substantive principles are also contained within the DoPs. Article III stipulates that the three countries shall, 'take all appropriate measures to prevent the causing of significant harm in utilising the Blue/Main Nile', and Article IV stipulates that they will 'utilise their shared water resources in their respective territories in an equitable and reasonable manner' (DoPs 2015). Notably, Article III also recognises that:

> where significant harm nevertheless is caused to one of the countries, the state whose use causes such harm shall, in the absence of agreement to such use, take all appropriate measures in consultations with the affected state to eliminate or mitigate such harm and, where appropriate, to discuss the question of compensation.
>
> (DoPs 2015: Art. III)

Article III, through the use of the term 'where nevertheless is caused', appears more in line with the UN Watercourses Convention and the CFA in suggesting that some level of harm may be tolerated, even if it is significant. This is substantiated by the suggestion that the mitigation of harm may be appropriate, without having to comply with a stronger requirement to eliminate such harm. However, unlike Article 5 of the CFA, and Article 7 of the UN Watercourses Convention, Article III of the DoPs falls short of explicitly linking significant harm to the principle of equity. This raises a question of how provisions addressing the same matter, should be dealt with. In this regard, the ILC has commented that, 'it is a generally accepted principle that when several norms bear on a single issue they should, to the extent possible, be interpreted so as to give rise to a single set of compatible obligations' (ILC 2006). It might therefore be concluded that while not explicit, Article IV should be read in tandem with Article III, and in favour of the recognition under customary international law that the principle of equitable and reasonable utilisation provides the overarching framework for determining distributive justice. Such an approach is substantiated by the list of factors that should be taken into account when determining what is equitable and reasonable, which are taken verbatim from the CFA. These factors include both the *effects* of the use or uses of the water resources in each Basin State (emphasis added) (DoPs 2015: Art. IV[d]).

A number of provisions of the DoPs then relate specifically to the GERD. In general terms, it is recognised that, 'the Purpose of GERD is for power generation, to contribute to economic development, promotion of transboundary cooperation and regional integration through generation of sustainable and reliable clean energy supply' (DoPs 2015: Art. II). The adoption of this article reflects an endorsement by all three countries of the regional significance and acceptance of GERD. Article II is followed by requirements whereby the three countries agree, upon the basis of joint studies, to develop guidelines and rules on the first filling of GERD and its annual operation (Art. V). The joint studies are to be facilitated through a Technical National Committee (TNC) of Egypt, Ethiopia and Sudan, 'to provide the necessary data and information in good faith and in a timely manner' (Art. VII). The three countries also agree, pursuant to Article V, to 'sustain cooperation and coordination on the annual operation of GERD with downstream reservoirs'. In addition, Ethiopia commits to, 'inform the downstream countries of any unforeseen or urgent circumstances requiring adjustments in the operation of GERD' (Art. V[c]).

Finally, the DoPs obligate the three countries, 'to settle disputes, arising out of the interpretation or implementation of this agreement, amicably through consultation or negotiation in accordance with the principle of good faith' (Art. X). If, however, the parties are unable to settle their dispute through consultation or negotiation, 'they may jointly request for conciliation, meditation, or refer the matter for consideration of the Heads of State/Heads of Government' (Art. X). While falling short of requiring arbitration or adjudication, this provision is in line with the more general requirement under international law that states should settle their disputes in a peaceful manner (Charter of the United Nations 1945: Art. 33).

As with the 1993 Framework, the legal status of the DoPs is not clear. The absence of formal procedures by which the states can demonstrate their consent to be bound by the instrument, for example, entry into force, ratification, and registration raises the uncertainty surrounding the instrument (Vienna Convention 1969: Art. 24). The countries have not ratified the DoPs, nor have they formally denounced the early agreements (Aman 2015). Moreover, the Egyptian Constitution is arguably at variance with the DoPs in committing the state to maintain, 'Egypt's historic rights' in the Nile River (Constitution of the Arab Republic of Egypt 2014: Art. 44).

Conversely, and as discussed above, ratification is not an essential criterion in determining whether or not an instrument is legally binding. Also, inconsistency with national laws does not preclude an international agreement being legally binding upon a state (see Vienna Convention 1969: Art. 27). Additionally, a state does not need to officially denounce prior legal commitments in order to endorse subsequent legal commitment (Vienna Convention 1969: Art. 59[1][b]). Such an abrogation of old Nile treaties based on the doctrine of *lex prior* can be justified and necessary for progressive development of the law (Klabbers 2011: 197). In terms of the text of the instrument, the DoPs certainly appears to have some normative character. Many of its provisions are couched in terms such as, 'the three countries shall' or 'will' (DoPs 2014: Arts II, IV, V, VI, VIII, IX, X). As discussed above, the DoPs also reflects endorsement by the countries of established principles of customary international law, such as equitable utilisation and no significant harm, and thus provides a rule-based procedural and substantive framework in the Eastern Nile Basin.

Fairness – what insights from the Blue Nile?

The study of the key cooperative agreements pertaining to the Blue Nile, as well as the law of international watercourses, demonstrates that there is considerable merit in focusing on questions of fairness. In particular, this study has shown the merits of coupling issues of legitimacy or 'right process', with questions of distributive justice, when considering legal regimes relating to international watercourses.

Form and Fairness

States enter into a wide range of cooperative agreements, with varying degrees of commitment, and for numerous different reasons (Guzman 2005). The choice of instrument has certain trade-offs. Raustiala, for instance, suggests that an 'overly deep' agreement, 'may lead to numerous violations, which could undermine future credibility or create political backlash against international cooperation' (Raustiala 2005: 613). Conversely, in the case of non-binding instruments, or what Raustiala calls 'pledges', 'states may negotiate deep pledges but do little or nothing to implement them' (Raustiala 2005: 611).

The study of the Nile illustrates the diverse range of instruments that states use in a bid to further cooperation. The colonial era treaties would appear to fit into the mode of a legally binding treaty and express normative commitment. Similarly, the CFA is an attempt to establish a basin-wide legally binding treaty. What, therefore, is the implication of choosing legally binding cooperative arrangements? Are they more likely to generate compliance? In the case of the Nile, it would appear that legitimacy issues have overshadowed questions of form. The early treaties are highly contested, primarily concerning matters of state succession, which are themselves disputed. Much work has been done by scholars in an attempt to ascertain the legality of these treaties (e.g. Woldetsadik 2013), but such analysis arguably misses a much more pertinent issue. From a legitimacy standpoint, it is clear that all states do not perceive these instruments to be fair, and their compliance pull is subsequently limited.

For different reasons, the compliance pull of the CFA is also a cause for concern. Contentious issues both in terms of its content, that is, Article 14(b), the process of its adoption, and subsequent limited number of ratifications are likely to influence its 'compliance pull'. However, subsequent developments both within and outwith the basin (as discussed below) may well shape the future significance of the CFA.

The form of the 1993 Framework and the DoPs is unclear. This in itself raises the question of whether ambiguity regarding the formal status of an instrument hinders compliance, or whether 'constructive ambiguity' might be seen as beneficial in furthering cooperation while avoiding any potential impasse in negotiations (Hafner 2013). There has been considerable discussion on the implications of 'constructive ambiguity' within treaty terms (Fishhendler 2008; Pehar 2001), and the difference between legally binding (hard law) and non-legally binding (soft law) instruments (e.g. Shelton 2000). However, less attention has been paid to the question of whether ambiguity as to the legal status of an instrument is in itself beneficial or not.

A rich avenue for further empirical research would be to explore the perception of states and other actors as to the legitimacy of the differing forms of cooperating arrangements that exist within the Nile, and other transboundary river basins, and consider what implications those perceptions have on the 'compliance pull' of such instruments. From a glance, it would appear that the ambiguity pertaining to the 1993 Framework agreement may have contributed to its poor 'compliance pull'. For the DoPs, it is too early to tell, although, despite some setbacks, initial signs would suggest that the states are taking their commitments thereunder seriously. Through subsequent meetings, the three countries have reaffirmed their commitment to the DoPs, and initiated the joint studies referred to in the instrument (Salman 2016: 524). Ultimately, the 'compliance pull' of the DoPs, and questions over form, might only be seriously tested when there are disagreements over its implementation and the way in which they are dealt with by both sides, or if the political will of any of the states to implement it changes.

Equity as distributive justice

An analysis of the key legal agreements relating to the Blue Nile has shown that there has been a gradual shift towards the notion of distributive justice within the basin. While the colonial era agreements were at odds with the notion of equity, the language of distributive justice has been incorporated in later agreements. Egypt and Sudan's reluctance, at least to date, to endorse the CFA and their insistence on the explicit mention of current uses within Article 14(b), might be seen as a blip in the move towards a clearer articulation of distributive justice within the Nile Basin. However, evidence of a positive shift towards the notion of distributive justice can be seen in the inclusion of the principle of equitable and utilisation within the NBI's vision, the CFA (Art. 4) and the DoPs (Art. IV). This trend no doubt enhances the potential 'compliance pull' of any cooperative arrangement that the Nile states enter into, and also demonstrates the inter-relationship and norm diffusion that takes place between cooperative arrangements created within different spaces and levels (Jacobs 2012). One might even conjecture that recognition of the key substantive norms of equitable and reasonable utilisation and no significant harm within the DoPs, soon after entry into force of the UN Watercourses Convention, was more than a coincidence.

This might, however, be criticised as a subjective application of distributive justice to a controversial project and subject matter. Given that distributive justice concerns the sharing of the costs and benefits of a transboundary resource, the full application of the notion of fairness can only be clearly and objectively determined once the terms of the DoPs are fully implemented, which will require further negotiations and agreements both on any benefit-sharing arrangements, for example, energy sharing, and on the filling and operating of the dam. Ethiopia is investing nearly US$5 billion dollars in the construction of the dam, with the expectation that, in return, it will gain from meeting its domestic electricity demand and earn hard currency from exporting energy (as opposed to agricultural products). As a matter of priority, Ethiopia also appears committed to selling electricity to Egypt and Sudan presumably at a reasonable price. Finding a fair price and an agreeable level of electricity supply to be sold from GERD to both Sudan and Egypt is, therefore, an important factor to any mutual determination of distributive justice in the Blue Nile. However, the uncertainty over the economic and environmental impacts of the GERD on Sudan and Egypt (Tawfik and Dombrowsky, Chapter 6), poses a key challenge to overcome in the pursuit of outcome that can be perceived as just to all.

Of course, the filling of the GERD may have a negative impact on Egypt. For example, Boehlert *et al.* (Chapter 7) find that as a result of rapid GERD filling, Egyptian hydropower and agricultural deliveries are expected to fall by 10.3 and 2.1 per cent respectively during the first three years. Similarly, Kahsay *et al.* (Chapter 8) find that GERD filling may result in a 6 to 18 per cent reduction of hydropower and an 11 per cent reduction in agricultural water supplies in Egypt. However, these figures are the worst-case scenarios and can only happen if the filling period takes place 'during a sequence of dry years' and Sudan increases irrigation water supplies (Kahsay *et al.*, Chapter 8).

Most importantly, the two chapters in this volume that reflect the data and findings of two renowned groups of economists agree that the aforementioned potential impacts to Egypt are likely to be negligible, when their implications to the Egyptian GDP is taken into account. Boehlert *et al.* (Chapter 7), for example, estimate that the overall impact of such reductions during the filling period is equivalent to 0.13 per cent of the Egyptian economy. This is primarily because the temporary reduction in Egyptian hydropower is negligible as

Egyptian hydropower only represents 8 per cent of the country's overall energy production.

The two economics studies in this volume further agree that after the GERD is filled, it will have basin-wide economic benefits – through energy provision and revenue generation to Ethiopia and Sudan, improved water availability and hence agriculture deliveries to Egypt and Sudan (Kahsay *et al.*, Chapter 8).

This contradicts what others have said about the impact of GERD. A 2013 'Cairo University Report' predicted a 'reduction in the water share of Egypt [as a result of GERD] will result in abandoning huge areas of agricultural lands and scattering millions of families' (Egyptian Chronicles 2013).

In contrast, the works of Boehlert *et al.* (Chapter 7) and Kahsay *et al.* (Chapter 8) conclude that the GERD will have a basin-wide economic benefits after filling. Kahsay *et al.* went to the extent that:

> As the GERD enters its operation phase, the basin-wide gain in real GDP due to the GERD operation rises to about US$2 billion, of which Ethiopia, Sudan and Egypt earn US$1,474 million, US$448 million and US$75 million respectively. Thus, the GERD enhances real GDP in all the Eastern Nile countries when it starts operating.
>
> (Kahsay *et al.* 2018: xxx[18])

However, they also stress that 'the distribution of the benefits is skewed with Ethiopia, Sudan and Egypt earning 74 per cent, 22 per cent and 4 per cent of the total basin-wide gain in real GDP, respectively' (Kahsay *et al.* 2018: xxx[18]), a finding which may be justified on the basis of considerations of historical injustice as considered earlier and the significant financial investment made on the GERD by Ethiopia. Alternatively, such a disparity in benefit-sharing may well need to be adjusted through appropriate trade-offs (Boehlert *et al.*, Chapter 7; Tawfik and Dombrowsky, Chapter 6).

The findings of the two economic chapters (Chapters 7 and 8) do need to be taken with caution as unequivocally conceded by the authors. First, dam filling will only be fair, if undertaken within the right hydraulic and climatic conditions (i.e. during wet and average rather than dry seasons) and within a short period of time (Chapters 7 and 8). In contrast, a reasonable period of time (say 6–7 years) (Kahsay *et al.*, Chapter 8) will be required to mitigate adverse impacts of filling downstream. Second, after the reservoir is filled with water, the dam operation should be coordinated with Sudanese and Egyptian dams, including the High Aswan Dam, to mitigate consequences and maximise benefits from the dam (Wheeler, Chapter 10). These aspects of cooperation have been included in the DoPs (Art. V), and as factors to be taken into account in the determination of distributive justice.

The potential for a fair solution to GERD as supported by the findings of the two groups of economists on the impacts of GERD after filling is broadly consistent with other major studies and reports on GERD such as the International, Non-Partisan Eastern Nile Working Group report on GERD (MIT Report

2014). The potential trade-offs to maximise benefits from GERD for all clearly exist (Boehlert *et al.*, Chapter 7).

A further area where a focus on distributive justice could be beneficial, as highlighted by Franck, is in relation to non-state actors. To date, much of the negotiations concerning distributive justice within the Nile have been conducted at a state–state level. The only reference to non-state actors within the treaty arrangements is contained in the CFA, which is not in force. There is, however, a growing trend to engage civil society actors in matters related to the management of the Nile Basin. An example of this can be seen in the establishment and support for the Nile Basin Discourse (NBD), a civil society network of over 850 members and partner organisations within the region (NBD 2011). Another specific example relevant to the GERD might be the visit to the GERD by Egyptian and Sudanese journalists (Tekle 2016), which may help enhance mutual understanding of the costs and benefits of the project among the public of the Nile Basin. As noted previously, human rights law might provide a useful platform by which to advance the role of non-state actors in processes related to the GERD that are focused on the determination of distributive justice (Bulto 2013).

The substance – process link

The importance of considering distributive justice alongside process (legitimacy) comes across clearly within a study of the Nile. Perceptions of a lack of legitimacy have constituted a major barrier to state compliance with early Nile treaties, irrespective of whether or not these treaties are legally binding. Failure to include all basin states within the negotiation of those instruments has been the main reason for their poor 'compliance pull'.

More emphasis on the processes by which legal arrangements are adopted, alongside the growing political dynamics of transboundary water cooperation, or 'hydropolitics' (Zeitoun *et al.* 2017), would be advantageous. In this regard, this paper has highlighted the central role that good faith plays when states are negotiating cooperative arrangements. Drawing upon the work of the ICJ and others, and taking into account hydropolitics, the developments in Blue Nile might help articulate what good faith requirements mean within a transboundary water context. The DoPs, for example, adopted good faith as one of its cardinal principles (Art. I), as a mode of exchange of data and information and facilitating joint studies among the three countries (Art. VII) and as a requirement of amicable dispute settlement (Art. X). But good faith has also been mentioned in the Declaration in relation to taking dam safety measures (Art. VIII) and can also be inferred from the commitment made to build confidence among the parties (Art. VI). A further area of research would therefore be to conduct empirical analyses of the negotiation processes related to Nile cooperative arrangements, with a view to developing key criteria for the interpretation of good faith in the negotiation of watercourse agreements.

The interrelationship between substance and process is also evident within the design of the cooperative arrangements. Drawing from the UN Watercourses Convention, the CFA contains the key procedural rules of the law relating to international watercourses. These procedural rules offer an important means by which to assess, and overtime, reassess issues of distributive justice within a basin. Some of these rules are contained within the DoPs, both implicitly and explicitly. There is, for instance, the due diligence requirement, 'to take all appropriate measures to prevent the causing of significant harm' (Art. III). As noted above, this general requirement might imply a series of requirements upon states, that include conducting an environmental impact assessment and engaging with stakeholders.

More explicitly, the DoPs contains a series of procedural requirements, such as the development of rules and guidelines for operating the GERD, informing downstream states of unforeseen or urgent circumstances, the establishment of 'an appropriate coordination mechanism, the exchange of data and information', and so on (DoPs 2015: Arts V and VII). The ultimate substantive objective of all these, in addition to ensuring reasonable use of Nile waters by the parties, is to ensure that the GERD contributes 'to economic development, promotion of transboundary cooperation and regional integration through generation of sustainable and reliable clean energy supply' (DoPs 2015: Art. II), which is key for fostering equity among the parties. Considering this as a central element of distributive justice may seem very broad and subjective, however, with closer cooperation and further studies on the subject, the project can promote and address the hopes and fears of all concerned.

While the design of the aforementioned instruments would suggest recognition of the importance of linking substance to process, exploring the nature of such a linkage necessitates further scholarly attention. To date, there has only been a general recognition of the correlation between substance and process. This has led some scholars to caution that elastic principles such as equity, may prove ineffective in the face of relative bargaining power, which in turn may disadvantage weaker states (Woodhouse and Zeitoun 2008: 103–119). This perceived blindness to bargaining power has been used to criticise Franck's heavy reliance on distributive justice (Scobbie 2002: 924). While under-explored, the observation was not lost on Franck, who recognised the role that 'an effective, credible, institutionalized, and legitimate interpreter' can play in providing an elastic rule within meaning within a particular case (Franck 1990: 81). Others have suggested that the role of interpreter should go further than what was envisaged by Franck, in order to encompass social processes and epistemic communities beyond intergovernmental platforms (Brunnée and Toope 2010: 52–53). Much work could be done to assess the extent to which existing treaty frameworks relating to international watercourse support such a broader process-based approach, and how those processes in turn affect interpretations of distributive justice – especially where power asymmetry influences cooperation between states. An effective regional or basin-wide approach to designing, interpreting and applying cooperative frameworks may well help mitigate, if not totally eliminate, the effects of power imbalance on equitable share of water resources. It is of note that Ethiopia, Sudan and Egypt jointly sponsored an international panel of experts to study of the downstream impacts of GERD (IPoE 2013) and are currently undertaking a follow-up third-party study involving two French firms (Salman, Chapter 3).

Conclusion

This paper has sought to explore the concept of fairness within the context of the law of international watercourses generally, the Nile Basin, and within the Blue Nile context. Exploring the concept of fairness within such contexts has offered up some interesting insights that speak to the question of why states comply with their international commitments. The study has shown that much can be gained by looking at the law relating to international watercourses as a case study by which to examine what fairness might actually mean in practice. This may be in terms of the design of cooperative arrangements, the determinacy of commitments, the interrelationship between substantive norms and process, the process by which cooperative arrangements are negotiated and the factors and interests that might be reconciled in the determination of equity. The case study has also shown the prospects and challenges of objectively determining the application of fairness 'on the ground'. It needs to be highlighted that the fairness principle is a framework, the actual implementation of which relating to specific areas of law, subject matter or a particular project is contingent on continuous cooperation and progressive realisation of the fair deal by the parties concerned, based upon the cardinal principles of good faith, mutual respect and equality. Therefore, in addressing distributive justice vis-à-vis the GERD it should be recognised that reaching agreement over every minute detail is not necessary nor advantageous. Rather, there is a need to – agree on certain principles and values that are perceived by the parties as substantively fair. However, the objective application and determination of fairness benefits from applying appropriate scientific studies that are crucial to fully appreciate the costs and benefits of an undertaking to all concerned and accordingly determine the distributive justice within a specific context. The interdisciplinary approach to the fairness principle is therefore beneficial to foster benefit sharing and identify appropriate trade-offs on a controversial subject of resource sharing among states, including on shared watercourses. Finally, it is shown that procedural fairness and distributive justice, as normative standards, constitute two sides of the same coin, which in turn, demands that both are taken considered together when analysing the merits of transboundary treaty frameworks.

References

Aal, M.S. (2013). Equitable and Reasonable Utilization of International Rivers in the UN Convention with a Particular Reference to the River Nile. *African Perspectives*, 11(39), 22–28.

Abdo, M. (2004). The Nile Question: The Accords on the Water of the Nile and Their Implications on Cooperative Schemes in the Basin. *Perceptions*, 45–57. Retrieved 20 May 2016, from http://sam.gov.tr/wp-content/uploads/2012/01/4.-Mohammed-Abdo.pdf.

Akehurst, M. (1976). Equity and General Principles of Law. *International and Comparative Law Quarterly*, 25(4), 801–825.

Aman, A. (2015). Egypt Warily Signs Preliminary Nile Agreement. *Almonitor*. Retrieved 21 May 2016, from www.al-monitor.com/pulse/originals/2015/03/egypt-eastern-nile-water-agreement-ethiopia-sudan.html.

Archer, J.L. (2012). Transcending Sovereignty: Locating Indigenous Peoples in Transboundary Water Law. Masters of Law Thesis, University of British Columbia. Retrieved 20 May 2016, from http://papers.ssrn.com/sol3/papers.cfm?abstract_id=1997539.

Boehlert, B., Strzepek, K.M. and Robinson, S. (2018). Analysing the Economy-Wide Impacts on Egypt of Alternative GERD Filling Policies. In Z. Yihdego, A. Rieu-Clarke and A.E. Cascão (eds), *The Grand Ethiopian Renaissance Dam and the Nile Basin: Implications for Transboundary Water Cooperation*. London: Routledge.

Boyle, A. and Chinkin, C. (2007). *The Making of International Law*. Oxford: Oxford University Press.

Brown-Weiss, E. (1990). In Fairness to Future Generations. *Environment: Science and Policy for Sustainable Development*, 32(3), 6–31.

Brunnée, J. and Toope, S.J. (2002). The Changing Nile Basin Regime: Does Law Matter. *Harvard International Law Journal*, 43(1), 105–159.

Brunnée J. and Toope, S.J. (2010). *Legitimacy and Legality in International Law: An Interactional Account*. Cambridge: Cambridge University Press.

Bulto, T.S. (2013). *The Extraterritorial Application of the Human Right to Water in Africa*. Cambridge: Cambridge University Press.

Cascão, A.E. and Nicol, A. (2018). Changing Cooperation Dynamics in the Nile Basin and the Role of the GERD. In Z. Yihdego, A. Rieu-Clarke and A.E. Cascão (eds), *The Grand Ethiopian Renaissance Dam and the Nile Basin: Implications for Transboundary Water Cooperation*. London: Routledge.

Case Concerning Pulp Mills on the River Uruguay (*Argentina v. Uruguay*) (2010). ICJ Reports of Judgements, Advisory Opinions and Orders, 20 April. Retrieved 20 May 2016, from www.icj-cij.org/docket/files/135/15877.pdf.

Chapman, J.W. (1963). Justice and Fairness. In C.J. Friedrich and J.W. Chapman (eds), *Nomos VI: Justice* (pp. 147–169). New York: Atherton Press.

Charter of the United Nations (1945). 26 June 1945, San Francisco. Retrieved 20 May 2016, from https://treaties.un.org/doc/publication/ctc/uncharter.pdf.

CFA (2010). Agreement on the Nile River Basin Cooperation Framework, open for signature 13 April 2010. Retrieved 21 May 2016, from www.nilebasin.org/index.php/spotlight/99-cfa-overview.

Constitution of the Arab Republic of Egypt (2014). 18 January. Retrieved 21 May 2016, from www.constituteproject.org/constitution/Egypt_2014.pdf.

Crawford, J. (2012). *Brownlie's Principles of Public International Law*, 8th Edn. Oxford: Oxford University Press.

D'Amato, A. (1992). Good Faith. In R. Bernhardt (ed.), *Encyclopaedia of Public International Law* (pp. 599–601). Amsterdam: North-Holland.

Degefu, G.T. (2003). *The Nile: Historical, Legal and Developmental Perspectives*. New York: Trafford Publishing.

Desierto, D.A. (2015). Rawlsian Fairness and International Arbitration. *University of Pennsylvania Journal of International Law*, 36(4), 939–993.

DoPs (2015). *Declaration of Principles between The Arab Republic of Egypt, the Federal Democratic Republic of Ethiopia and the Republic of the Sudan on the Grand Ethiopian Renaissance Dam Project* (GERDP). *Horn Affairs*, 25 March. Retrieved 20 May 2016, from http://hornaffairs.com/en/2015/03/25/egypt-ethiopia-sudan-agreement-on-declaration-of-principles-full-text/.

Egyptian Chronicles (2013, 17 June). Cairo University's report on Ethiopia's Great Renaissance Dam. Retrieved 10 May 2017, from http://web.mit.edu/12.000/www/m2017/pdfs/ethiopia/cairo.pdf.

Egyptian Ministry of Foreign Affairs (n.d.). Egypt's Perspective Towards the Ethiopian Grand Renaissance Dam Project (GERDP). Retrieved 21 May 2016, from www.mfa.gov.eg/English/EgyptianForeignPolicy/Pages/renaissance_dam.aspx.

Fishhendler, I. (2008). When Ambiguity in Treaty Design Becomes Destructive: A Study of Transboundary Water. *Global Environmental Politics*, 8(1), 111–136.

Framework for General Cooperation Between The Arab Republic of Egypt and Ethiopia, Cairo (1993). 1 July. Retrieved 20 May 2016, from http://gis.nacse.org/tfdd/tfdddocs/521ENG.pdf.

Franck, T. (1988). Legitimacy in the International System. *American Journal of International Law*, 82(4), 705–759.

Franck, T. (1990). *The Power of Legitimacy Among Nations*. Oxford: Oxford University Press.

Franck, T. (1995). *Fairness in International Law and Institutions*. Oxford: Clarendon Press.

Fuentes, X. (1996). The Criteria for the Equitable Utilisation of International Rivers. *British Yearbook of International Law*, 67(1), 337–412.

Garane, A. and Abdul-Kareem, T. (2013). West Africa. In F. Loures and A. Rieu-Clarke (eds), *The UN Watercourses Convention in Force: Strengthening International Law for Transboundary Water Management* (pp. 97–111). London: Earthscan.

Garner, B. (1999). *Black's Law Dictionary*, 7th Edn. St Paul, MN: West Publishing.

Garretson, A.H. (1967). The Nile Basin. In A.H. Garretson, R.D. Hayton and C.J. Olmstead (eds), *The Law of International Drainage Basins* (pp. 256–297). Dobbs Ferry, NY: Oceana Publications.

Guzman, A.T. (2005). The Design of International Agreements. *European Journal of International Law*, 16(4), 579–612.

Hafner, G. (2013). Subsequent Agreements and Practice: Between Interpretation, Informal Modification, and Formal Amendment, In G. Nolte (ed.), *Treaties and Subsequent Practice*. Oxford: Oxford University Press.

Hsu, S. (2004). Fairness versus Efficiency in Environmental Law. *Ecology Law Quarterly*, 31(2), 303–402.

ILC (1994). Draft Articles on the Law of the Non-navigational Uses of International Watercourses and commentaries. *Yearbook of the International Law Commission*, II(2), 89–135.

ILC (2006). Fragmentation of International Law: Difficulties Arising from the Diversification and Expansion of International Law, UN Doc. A/CN.4/L.672, 13 April 2006. Retrieved 20 May 2016, from http://legal.un.org/ilc/documentation/english/a_cn4_1682.pdf.

International Panel of Experts (IPoE) (2013). *Grand Ethiopian Renaissance Dam: Final Report*.

Jacobs, I.M. (2012). *The Politics of Water in Africa: Norms, Environmental Regions and Transboundary Cooperation in the Orange-Sengu and Nile Rivers*. London: Continuum.

Jennings, R. (1986). Equity an Equidistance Principle. *Annuaire. Suisse de Droit International*, XLII, 27–38.

Kahsay, T.N., Kuik, O., Brouwer, R. and Van der Zaag, P. (2018). Economic Impact Assessment of the Grand Ethiopian Renaissance Dam under Different Climate and Hydrological Conditions. In Z. Yihdego, A. Rieu-Clarke and A.E. Cascão (eds), *The Grand Ethiopian Renaissance Dam and the Nile Basin: Implications for Transboundary Water Cooperation*. London: Routledge.

Kaya, I. (2003). *Equitable Utilisation: The Law of Non-Navigational Uses of International Watercourses*. Aldershot: Ashgate.

Kimenyi, M. and Mbaku, J. (2015). *Governing the Nile River Basin: The Search for a New Legal Regime*. Arlington: Brookings Institution Press.

Klabbers, J. (2011). The Vienna Convention and Conflicting Treaty Provisions. In Enzo E. Cannizzaro and M.H. Arsanjani (eds), *The Law of Treaties Beyond the Vienna Convention* (pp. 192–205). Oxford: Oxford University Press.

Lake Lanoux Arbitration (*France v. Spain*). (1957). Reproduced in *International Law Reports*, 24, 101.

Leb, C. (2012). The Right to Water in a Transboundary Context: Emergence of Seminal Trends. *Water International*, 37(6): 640–653.

Leb, C. (2013). *Cooperation in the Law of Transboundary Water Resources*. Cambridge: Cambridge University Press.

Leb, C. and Tignino, M. (2013). State Succession to Water Treaties: Uncertainties and Extremes. In L.B. de Chazournes, C. Leb and M. Mara Tignino (eds), *International Law and Freshwater: The Multiple Challenges* (pp. 421–444). Cheltenham: Edward Elgar.

Legal Consequences of the Construction of a Wall in the Occupied Palestinian Territory, Separate Opinion of Judge Owada (2004). ICJ Reports of Judgements, Advisory Opinions and Orders, 9 July (pp. 260–271). Retrieved 19 May 2016, from www.icj-cij.org/docket/files/131/1691.pdf.

Liguori, T. (2009). The Principle of Good Faith in the Argentina–Uruguay Pulp Mills Dispute. *Journal of Water Law*, 20(2/3): 70–75.

Lipper, J. (1967). Equitable Utilisation. In A.H. Garretson, R.D. Hayton and C.J. Omstead (eds), *The Law of International Drainage Basins* (pp. 15–88). Dobbs Ferry, NY: Oceana Publications.

Louka, E. (2006). *International Environmental Law: Fairness, Effectiveness and World Order*. Cambridge: Cambridge University Press.

McCaffrey, O. (2007). *The Law of International Watercourses*, 2nd Edn. Oxford: Oxford University Press.

McCaffrey, S. (1992). Human Right to Water: Domestic and International Implications. *Georgetown International Environmental Law Review*, 5: 1–24.

McIntyre, O. (2010). The Proceduralisation and Growing Maturity of International Water Law. *Journal of Environmental Law*, 22(3): 475–495.

McIntyre, O. (2013). Utilization of Shared International Freshwater Resources: The Meaning and Role of 'Equity' in International Water Law. *Water International*, 38(2): 112–129.

McKenzie, S.O. (2012). Egypt's Choice: From the Nile Basin Treaty to the Cooperative Framework Agreement, an International Legal Analysis. *Transnational Law & Contemporary Problems*, 21. Retrieved 20 May 2016, from http://papers.ssrn.com/sol3/papers.cfm?abstract_id=2445962##.

Maluwa, T. (1999). *International Law in Post-Colonial Africa*. Leiden: Martinus Nijhoff.

Martens, E. (2012). Article 102. In B. Simma, D.-E. Khan, G. Nolte and A. Paulus (eds), *The United Nations Charter: Commentary*, 3rd Edn (pp. 2098–2109). Oxford: Oxford University Press.

Mekonnen, D. (2010). The Nile Basin Cooperative Framework Agreement Negotiations and the Adoption of a 'Water Security' Paradigm: Flight into Obscurity or a Logical Cul-de-Sac? *European Journal of International Law*, 21(2), 421–440.

MIT (2014). The Grand Ethiopian Renaissance Dam: An Opportunity for Collaboration and Shared Benefits in the Eastern Nile Basin. Retrieved 1 June 2016, from http://jwafs.mit.edu/sites/default/files/documents/GERD_2014_Full_Report.pdf.

Mitchell, L.E. (1993). Fairness and Trust in Corporate Law. *Duke Law Journal*, 43(3), 425–491.

NBD (2011). Civil Society Engagement in Nile Cooperation and Development Project. Annual Report. Retrieved 22 May 2016, from http://nilebasindiscourse.org/images/downloads/NBD-Annual-Report-2011.pdf.

NBI (2016a). The Cooperative Framework Agreement for the Nile Basin: An overview. Retrieved 21 May 2016, from www.nilebasin.org/index.php/spotlight/99-cfa-overview.

NBI (2016b). Nile Basin Initiative. Retrieved 20 May 2016, from www.nilebasin.org/index.php/about-us/nile-basin-initiative.

Nile Treaty (1902). Treaty between the United Kingdom and Ethiopia, Addis Ababa, 15 May. Retrieved 20 May 2016, from http://treaties.fco.gov.uk/docs/pdf/1902/TS0016.pdf.

Nile Treaty (1929). Exchange of Notes between Her Majesty's Government in the United Kingdom and the Egyptian Government on the Use of Waters of the Nile for Irrigation, Cairo, 7 May. Retrieved 20 May 2016, from www.internationalwaterlaw.org/documents/regionaldocs/Egypt_UK_Nile_Agreement-1929.html.

Nile Treaty (1959). Agreement between the Republic of the Sudan and the United Arab Republic for the Full Utilisation of the Nile Waters, Cairo, 8 November 1959 (entered into force 12 December 1959). *Treaty Series*, XCIII, p. 43.

North Sea Continental Shelf Cases (*Federal Republic of Germany/Denmark; Federal Republic of Germany/Netherlands*) (1969). ICJ Reports Judgements, Advisory Opinions and Orders, 20 February. Retrieved 20 May 2016, from www.icj-cij.org/docket/files/52/5561.pdf.

O'Connor, J.F. (1991). *Good Faith in International Law*. Brookfield, VT: Dartmouth Publishing Co.

Patrick, M.J. (2014). The Cycles and Spirals of Justice in Water-Allocation Decision Making. *Water International*, 39(1), 63–80.

Pehar, D. (2001). Use of Ambiguities in Peace Agreements. In J. Kurbalija and H. Slavik (eds), *Language and Diplomacy* (pp. 163–200). Msida, Malta: DiploProjects, Mediterranean Academy of Diplomatic Studies.

Raustiala, K. (2005). Form and Substance in International Agreements. *American Journal of International Law*, 99, 581–614.

Rawls, J. (1958). Justice as Fairness. *The Philosophical Review*, 67(2), 164–194.

Rawls, J. (1971). *The Law of Peoples: The Idea of Public Reason Revisited*. Boston, MA: Harvard University Press.

Salman, S.M.A. (2004). *The Human Right to Water: Legal and Policy Dimensions*. Washington, DC: World Bank.

Salman, S.M.A. (2013). The Nile Basin Cooperative Framework Agreement: A Peacefully Unfolding African Spring?. *Water International*, 38(1), 17–29.

Salman, S.M.A. (2016). The Grand Ethiopian Renaissance Dam: The Road to the Declaration of Principles and the Khartoum Document. *Water International*, 41(4),

512–527. Retrieved 22 May 2016, from www.tandfonline.com/doi/pdf/10.1080/025080 60.2016.1170374?instName=University+of+Dundee.

Salman, S.M.A. (2018a). The Nile Basin Cooperative Framework Agreement: Disentangling the Gordian Knot. In Z. Yihdego, A. Rieu-Clarke and A.E. Cascão (eds), *The Grand Ethiopian Renaissance Dam and the Nile Basin: Implications for Transboundary Water Cooperation*. London: Routledge.

Salman, S.M.A. (2018b). Agreement on Declaration of Principles on the GERD: Levelling the Playing Field. In Z. Yihdego, A. Rieu-Clarke and A.E. Cascão (eds), *The Grand Ethiopian Renaissance Dam and the Nile Basin: Implications for Transboundary Water Cooperation*. London: Routledge.

Sarvarian, A., Fontanelli, F., Baker, R. and Tzevelekos, V. (2015). *Procedural Fairness in International Courts*. London: British Institute of International and Comparative Law.

Scobbie, I. (2002). Tom Franck's Fairness. *European Journal of International Law*, 13(4), 909–925.

Shaw, M. (1996). The Heritage of States: the Principle of *uti possidetis juris* Today. *British Yearbook of International Law*, 67(1), 75–154.

Shelton, D. (ed.) (2000). *Commitment and Compliance: The Role of Non-Binding Norms in the International Legal System*. Oxford: Oxford University Press.

Shetewy, M.A. (2013). Legal Commitments regulating the Establishment of Water Projects on International Rivers Application Study over the Nile Basin. *African Perspectives*, 11(39), 34–35.

Soltau, F. (2009). *Fairness in International Climate Change Law and Policy*. Cambridge: Cambridge University Press.

Tawfik, R., and Dombrowsky, I. (2018). GERD and Hydropolitics in the Eastern Nile: from Water-Sharing to Benefit-Sharing?. In Z. Yihdego, A. Rieu-Clarke and A.E. Cascão (eds), *The Grand Ethiopian Renaissance Dam and the Nile Basin: Implications for Transboundary Water Cooperation*. London: Routledge.

Tekle, T-A. (2016, 2 August). Nile Basin Journalists Trained on Water Issues. *Sudan Tribune*. Retrieved 8 May 2017, from www.sudantribune.com/spip.php?iframe&page=imprimable&id_article=59809.

UN CESCR (2002). *General Comment No. 15 (2002) – The Right to Water (Arts 11 and 12 of the International Covenant on Economic, Social and Cultural Rights)*, UN Doc. E/C.12/2002/11, 20 January 2003. Retrieved 20 May 2016, from http://unhcr. org/49d095742.html.

UNECE WC (1992). *Convention on the Protection and Use of Transboundary Watercourses and International Lakes*, Helsinki, 17 March 1992 (entered into force 6 October 1996), *International Legal Materials*, 31, 1312.

UN GA Res (2010). *The Human Right to Water*. UN General Assembly Resolution 64/292, 3 August 2010, Retrieved 20 May 2016, from www.un.org/es/comun/ docs/?symbol=A/RES/64/292&lang=E.

UNWC (1997). *Convention on the Law of the Non-navigational Uses of International Watercourses*, New York, 21 May 1997 (entered into force 17 August 2014), *International Legal Materials*, 36, 700.

Vienna Convention (1969). *Vienna Convention on the Law of Treaties*. Vienna, 23 May 1969. *UN Treaty Series*, 1155, 331.

Vienna Convention (1978). *Vienna Convention on the Succession of States in Respect of Treaties*. Vienna, 23 August 1978 (entered into force 6 November 1996). *UN Treaty Series*, 1946, 3.

Wheeler, K.G. (2018). 'Managing Risks While Filling the Grand Ethiopian Renaissance Dam. In Z. Yihdego, A. Rieu-Clarke and A.E. Cascão (eds), *The Grand Ethiopian Renaissance Dam and the Nile Basin: Implications for Transboundary Water Cooperation*. London: Routledge.

Woldetsadik, T.W. (2013). *International Watercourses Law in the Nile River Basin: Three States at a Crossroads*. Abingdon: Routledge.

Woodhouse, M. and Zeitoun, M. (2008). Hydro-Hegemony and International Water Law: Grappling with the Gaps of Power and Law. *Water Policy, 10*(2), 103–119.

Wouters, P. and Chen, H. (2013). China's 'Soft-Path' to Transboundary Water Cooperation Examined in the Light of Two UN Global Water Conventions – Exploring the 'Chinese Way'. *Journal of Water Law, 22*, 229–247.

Zeitoun, M., Cascão, A.E., Warner, J., Mirumachi, N., Matthews, N., Menga, F. and Farnum, R. (2017). Transboundary Water Interaction III: Contest and Compliance. *International Environmental Agreements: Politics, Law and Economics, 17*(2), pp. 1–14. Retrieved 22 May 2016, from http://link.springer.com/article/10.1007/s10784-016-9325-x.

Ziganshina, D. (2013). Central Asia. In F. Loures and A. Rieu-Clarke (eds), *The UN Watercourses Convention in Force: Strengthening International Law for Transboundary Water Management* (pp. 152–167). London: Earthscan.

Ziganshina, D. (2015). Promoting Transboundary Water Security in the Aral Sea Basin Through International Law. Leiden: Koninklijke Brill NV.

5 Changing cooperation dynamics in the Nile Basin and the role of the GERD

Ana Elisa Cascão and Alan Nicol

Introduction

In recent years much has been discussed about the impacts the Grand Ethiopian Renaissance Dam (GERD) is having or will have on hydropolitical relations between Nile riparian states and the wider cooperation process (Bayeh 2015; Gebreluel 2014; Whittington *et al.* 2014; Tawfik 2015, 2016). These analyses usually frame the GERD as a major game changer, representing the beginning of a new era in the Nile Basin, and focus mainly on the political impacts of the GERD. The authors of this article contend, however, that before analysing the GERD as the catalyst of change, we need to start by understanding the GERD as an outcome of change. In particular, there is a need to understand the GERD as a direct outcome of a shifting and complex multilateral cooperation process that began in the mid-1990s. This article comprises two separate (though complementary) parts: the first part analyses the GERD as an outcome of change, while the second analyses the GERD as a cause of further change. Overall, the analysis is informed by critical theoretical concepts such as those under the 'Transboundary Water Interaction' framework, developed by the London Water Research Group (Zeitoun and Mirumachi 2008; Zeitoun, Mirumachi and Warner 2011; Zeitoun *et al.* 2014, 2017).

The first part of this article analyses how the GERD can be understood as an outcome of change. Going back to early 2010, the scenario was one of strengthening multilateral cooperation, against a background of past hydropolitical conflict and mistrust. From 1999 onwards, under the Nile Basin Initiative (NBI), the Nile riparians had sought the achievement of a shared vision for the Basin. This cooperation included joint identification, study and planning of investment projects that would, it was envisaged, bring tangible benefits to Nile countries and their populations. For example, the NBI developed a detailed portfolio of investment projects on energy production and trade, agriculture, watershed management and environmental protection (NBI 2014). In the hydropolitically complex Eastern Nile Basin, the three riparian states worked together under the Eastern Nile Technical Regional Office (ENTRO) towards the preparation of the ambitious Joint Multipurpose Project (JMP), which envisaged development of hydraulic infrastructure in the Blue Nile Basin. In this context, the article attempts to analyse Ethiopia's decision to move ahead in 2011 with the GERD as a national project, when it was originally expected that a series of dams would be built on the Blue Nile under the NBI, and ENTRO's JMP in particular. Two questions are addressed: (1) can we assume that the GERD is an outcome of failure and/or delays in multilateral cooperation?; or (2) should we assume that the GERD is actually an indirect (though unintended) consequence of the evolution of a wider cooperation process? This article seeks to understand how the 'cooperation process and norms' have actually contributed to changing the *status quo* in the basin and placed the GERD as a major new factor within the wider hydropolitical landscape.

The second part of this article looks at GERD as a catalyst of further change and a shaper of future developments and cooperation. It will discuss how the announcement of GERD's construction in 2011 was followed by technical talks and cooperation – and then by political negotiations – even to the extent of entailing the adoption of a new (trilateral) legal instrument in March 2015. In brief, the GERD is a milestone in the Eastern Nile landscape from several perspectives: not only in hydropolitical terms, but also in terms of regime change over the management of the shared water resources, the dynamics of water utilisation and management, economics and incentives for regional economic integration approaches, and a generalised awareness that water cooperation is more essential than ever. These new norms and processes will be analysed, as well as their hydropolitical impacts. Finally, the article looks at what wider implications the new hydropolitical landscape in the Eastern Nile might have on Nile Basin multilateral cooperation, namely, under the NBI, ENTRO and Cooperative Framework Agreement (CFA) processes and the eventual establishment of a Nile Basin Commission.

Pre-GERD cooperation norms and processes

At the beginning of 2011, when the Ethiopian government publicly announced its intention to build the GERD, the Nile Basin cooperation process was already in a state of flux and the region was experiencing a set of new and unprecedented socio-political, economic and hydropolitical dynamics (Nicol and Cascão 2011; Cascão and Nicol 2013; Matthews *et al.* 2013). This first section analyses how, from the 1990s onwards, the Nile riparians had been working together towards the establishment and strengthening of cooperative institutions, and how new cooperative norms were being adopted. Further, it analyses how political events in 2010 – when the signing of the Nile CFA by upstream riparians took place, with consequent reactions by downstream riparians – affected the ongoing cooperation process. The section also briefly analyses the relevant political changes at the national level in the three Eastern Nile countries that are also considered to have had hydropolitical impacts. This section introduces the key analytical elements for discussion in the next section, namely, how the GERD – and Ethiopia's decision to build the dam – can actually be considered to be an outcome of changes (and not only a catalyst of change, as most authors believe).

1990s: towards a new multilateral cooperative setting

In the 1990s, all the Nile riparian countries came together to establish a multi-lateral initiative to manage Nile Basin water resources. The main goal was to move from a legacy of hydropolitical conflict towards one of regional collaboration and cooperation. This process was strongly supported not only by the riparian countries themselves but also by the international community, including bilateral and multilateral development partners (Nicol et al. 2001; GTZ 2007). Until the 1990s, the Nile region was mainly characterised by an asymmetric development of water resources and a lingering diplomatic hydropolitical conflict, which was intimately linked to different political positions on past legal agreements regarding allocation of the Nile waters (Brunnée and Toope 2002). On the one hand, Egypt and Sudan had been the main users of the Nile, with hydraulic infrastructure in place and a bilateral agreement signed in 1959 that defined their rights and specific water quotas (Agreement 1959). On the other hand, the upstream riparians had hitherto made little use of Nile river water resources, had limited hydraulic infrastructure and had no agreements defining their own water rights, while highly contesting the 1959 Agreement (Arsano and Tamrat 2005). Up to the mid-1990s, the scenario was one of a prevailing and long-standing hydro-hegemonic stasis – downstream riparians wielding stronger material, bargaining and ideational power capable of influencing the outcomes of hydropolitical interaction across the basin and maintaining their hegemonic position; and upstream riparians exhibiting weak capacity to change this status quo due to intrinsic and extrinsic factors, including low economic development, limited capacity to exert power at regional and international levels and a global political scene that had been marked up until 1989 by relative stasis under the Cold War (Cascão 2009; Cascão and Zeitoun 2010a, 2010b; see also Zeitoun et al. 2011). It was in this context that past attempts to promote multilateral cooperation had been unsuccessful, partly because they would not address the major challenge, namely, the need for a new multilateral framework agreement (Brunnée and Toope 2002; Arsano and Tamrat 2005).

The new cooperation process, initiated in the mid-1990s, brought a novel approach to promoting transboundary cooperation under two parallel tracks: a technical track, with the Nile Basin Initiative (NBI) as a transitional cooperative arrangement; and a political track, driving negotiations for a CFA, representing a new legal and institutional framework. The rationale was that once countries adopted the CFA, the NBI as a transitional arrangement would be replaced by a permanent river basin commission (Brunnée and Toope 2002; NBI 2002). Meanwhile, the countries were expected to build progressively new cooperative norms through joint activities under a Shared Vision Program (SVP) and two Subsidiary Action Programmes (SAPs). In the ensuing decade, much was achieved. For example, under the SVP, the NBI promoted collaborative action, exchange of experience, and trust and capacity building, the main goal of which was the creation of an enabling environment for investments and action on the ground (NBI 1999, 2009; Cascão 2012). Later in the process, as an outcome of the SVP, the NBI would be entrusted with three core functions: water resources management, water resources development, and promotion of basin cooperation (NBI 2014). However, it was under the two SAPs – Eastern Nile Subsidiary Action Program (ENSAP) and Nile Equatorial Lakes Subsidiary Action Program (NELSAP) – that major normative changes took place in terms of thinking (and action) on water resources management and development at a transboundary level. The goals of the SAPs were to identify cooperative investment projects at the sub-basin levels that would confer mutual benefits to riparians and contribute to realizing transboundary development projects on the ground (NBI 1999). ENSAP and NELSAP, through multiple projects, promoted the joint identification, study and planning of hydraulic projects (including large-scale infrastructure) that would bring tangible benefits to the countries in the Eastern and Equatorial Nile sub-basins, respectively. The SAPs developed an impressive portfolio of investment projects with potential to deliver socio-economic benefits in the fields of energy, agriculture and environmental protection, etc. (NBI 2009, 2014). For the first time in the history of the Nile Basin, upstream and downstream countries established a joint and all-inclusive platform through which to discuss, consult and even implement optimal approaches to the development of shared water resources. It was under ENSAP/ENTRO, for example, that Ethiopia, Egypt and Sudan agreed upon and jointly implemented eight region-based projects (ENSAP 2009). Next, we look at the particular case of the ENSAP JMP to show how these new cooperative norms and processes worked in practical terms.

Joint Multipurpose Project: a promising example of a 'new set' of cooperation norms in the Eastern Nile Basin

The Joint Multipurpose Project (JMP) was one of eight ENSAP projects agreed upon by the Eastern Nile countries in 2003. Most of the projects were fast-track, small-scale and sectoral projects. In contrast, however, the JMP was a long-term, large-scale and multipurpose investment project aimed at identifying major optimal outcomes in terms of water resources management for all three Eastern Nile sub-basins (Blue Nile, Atbara/Tekeze and Baro-Akobo-Sobat). In 2005, the Eastern Nile Council of Ministers (ENCOM) started the launch phase of the JMP with the objective to: 'identify and prepare a major initial project, within a broader multipurpose programme, to demonstrate the benefits of a cooperative approach to the management and development of the Eastern Nile' (World Bank 2009). The JMP would build on earlier cooperative efforts already endorsed by the three countries under ENTRO: the Power Trade Study; an associated pre-feasibility study of three hydropower sites on the Blue Nile; and an ongoing feasibility study of regional power transmission interconnections. In 2007, a series of high-level meetings was held, leading to a joint call to accelerate cooperative action on the ground (ENSAP 2007). As a follow-up, the ENCOM commissioned an independent scoping study on the 'Opportunities for Cooperative Water Resources Development on the Eastern Nile: Risks and

Rewards' (Blackmore and Whittington 2008). This study represented an important milestone and landmark for Nile Basin cooperation. It was the first time that such a significant study had been jointly commissioned, and it was expected that such ground-breaking work could contribute to crucial changes. The high-level political involvement exhibited by all three countries provided evidence of expectations levels for the project, and also showed how the normative focus had begun shifting towards joint management and development. High expectations existed that the multi-billion-dollar JMP would become the first large-scale project in the Eastern Nile Basin to apply a genuinely transboundary approach by including: (1) planning and implementation based on regional decision-making processes; (2) regional confidence building based on joint communication and consultation mechanisms; (3) a benefit-cost-sharing formula under a 'no-borders perspective'; and (4) that it would ultimately contribute to ensuring efficient and optimal use of the Nile waters (Eldaw and Fekade 2009).

The JMP's long-term objective was to provide a range of transformational development benefits across sectors and countries (World Bank 2009). It was anticipated that it would unlock further linkages in terms of regional cooperation, trade and integration, and provide transformational socioeconomic benefits to the region. Of key relevance to this article is the fact that the JMP scoping study concluded that the Blue Nile sub-basin in Ethiopia provided the best opportunity for a first set of investments, including new water storage facilities that could generate large amounts of hydropower, and that this would provide important multipurpose benefits to downstream riparians, including flood control, sediment management and water availability for irrigation, etc. (Blackmore and Whittington 2008). This launching phase of the JMP was expected to be followed by a further phase under which countries would identify the first projects to be implemented and then move to implementation, while continuing to build the enabling technical and political environment for a new transformational approach to managing the Blue Nile waters (Cascão 2012). Interviews by the authors with relevant key authorities in Cairo, Addis and Khartoum (in 2008 and 2009) revealed that soon after the Scoping Study was released, the Egyptian authorities contested the findings and challenged the technical validity of the study conducted by the independent experts, following which their involvement in the JMP started to wane. In 2008 and 2009, there were several attempts by Ethiopia and Sudan through ENSAP to bring Egypt fully back into the JMP. Implicitly, it was assumed that the joint development of the Blue Nile Basin was the best-case scenario for all three countries, in particular when compared to the likely consequences of untrammelled unilateral developments.

However, in 2010, due to disagreements within the political cooperation track (see next section), the expectation that the JMP project could still offer a ground-breaking platform for joint large-scale hydraulic development evaporated, and the project came to a halt in 2012, before it could deliver any of the anticipated tangible results. In brief, the JMP provided clear evidence of a 'window of opportunity' for Eastern Nile countries: if major long-standing political stumbling blocks could be overcome, countries would have to eventually agree and jointly implement large-scale projects in the Blue Nile, and later in other Nile sub-basins. However, this window of opportunity rapidly shut.

Political track: progress and failure to reach a multilateral framework agreement

While the Nile riparian countries and NBI institutions were expanding transboundary technical cooperation, and at the same time imprinting a new set of cooperation norms, under the parallel legal track negotiators from all Nile riparians worked towards the achievement of a multilateral agreement – the CFA. Negotiations had been initiated in 1997, the goal of which was to agree on an institutional mechanism for cooperation among the Nile Basin states. The first part of the agreement (Agreement 2010) dealt with guiding legal principles; the second part provided for more specific principles regarding the institutional structure of a Nile River Basin Commission. Once the CFA was agreed upon, ratified and adopted, this Commission would replace the transitional arrangement – the NBI (NBI 2002).

From 1997 to 2007, the negotiations experienced several different phases and there was general optimism regarding outcomes (Brunnée and Toope 2002; Salman 2013). The Nile Basin was experiencing a new type of setting, which involved all riparians in open discussion on key transboundary issues and, ostensibly, in common search for a comprehensive agreement. Up until that point, negotiations between legal and political representatives of all Nile riparian countries had led to agreement on all forty-five legal articles except Article 14b, on water security (see Annex of Agreement 2010). Egypt and Sudan, however, had expressed strong reservations on Article 14b, because of possible implications on their current utilisation of the Nile waters and what they considered their 'historical and acquired rights', as enshrined in the 1959 Agreement (Reuters 2009; State Information System, Egypt 2011). This argument was similar to the long-standing official Egyptian position on any new legal agreement in the basin (Amer 1997). In 2007, confronted with a negotiation deadlock, the countries decided to refer the issue upwards to heads of state. However, the result was a stalemate that lasted until May 2009, at which point, during a Nile Council of Ministers meeting in Kinshasa, all upstream riparians decided that they would not wait any longer (The East African 2009). Instead, they decided to annex the controversial Article 14b and press on with signing the remaining negotiated articles. The signature by six countries occurred on 14 May 2010. As of early 2017, three countries had signed and ratified (Ethiopia, Rwanda and Tanzania), and four others were in the process of ratification/accession (Kenya, Uganda, Burundi and South Sudan).

The fact that a coalition of upstream countries had decided to sign the CFA and proceed to ratification, rejecting pressure from both Egypt and Sudan, represented a major tipping point in Nile hydropolitical relations, both in hydropolitical and legal terms – in the sense that upstream countries openly challenged the downstream position and mounted an effective joint upstream challenge to the legal *status quo*, that is, the hydro-hegemonic setting that had been in place in the Nile Basin for several decades (Zeitoun *et al.* 2014, 2017). However, this unprecedented move by upstream countries came with consequences. As a reaction to the CFA signature, in June 2010, both Sudan and Egypt froze their participation in all NBI activities and projects (Sudan Tribune 2010). This immediately contributed to slowing down the implementation of ongoing NBI projects and hindered the mobilisation of external funding, which implicitly contributed to problems of institutional and financial sustainability (NBI 2014). Sudan later changed its position (in November 2012) and returned as a full member of the NBI and ENTRO, after considering that there was no alternative to basin-wide cooperation between all Nile riparians. Subsequently, Sudan has used its leveraging powers to try to coax Egypt back into the NBI, though so far without success (Sudan Tribune 2012a; AFP 2014). The fact is that almost two decades after the transboundary cooperation process was initiated, the NBI continues to be a transitional arrangement experiencing significant challenges and the process of adopting a comprehensive legal framework has yet to be finalised. However, the Nile institutions and riparians, including Egypt, continue to highlight that multilateral cooperation is deemed necessary in order to promote regional economic development and integration, and – more implicitly – regional peace and stability (NBI 2016; SIS 2016).

National economic and political changes with regional hydropolitical impacts

Although this article aims mainly to analyse the regional dimensions (and associated changes) that have contributed to GERD becoming a reality on the ground, the fact is that one cannot ignore other significant and concomitant economic and political changes at the national level in all three Eastern Nile countries that have contributed to this new hydropolitical scenario. On the one hand, Ethiopia, a country marred by civil war up to 1991 amid deep poverty and food insecurity, has changed significantly in the last two decades. Improved political stability, strong leadership by the late Prime Minister Meles Zenawi, new economic and trade partnerships, and substantial economic growth have transformed the country into an icon for development in Sub-Saharan Africa with increasing demand for infrastructure and food and energy resources, while, at the same time, displaying growing capacity to mobilise funds domestically for the implementation of large-scale projects. Numbers show that Ethiopia's GDP has been increasing at around 10 per cent per annum on average in the last decade, and that the government has dramatically expanded national infrastructure in order to meet increasing national economic demands (AfDB, OECD and UNDP 2015). This has also been supported by a substantial increase in foreign direct investment in the country (Financial Times 2015). These factors have contributed to increasing the ambition and capacity of the Ethiopian government to project economic as well as political power, both within its borders and regionally. In the last decade, Ethiopia has signed several economic and trade agreements with its East African neighbours, including Kenya and Sudan, including on power trade (Makonnen and Lulie 2014; Verhoeven 2011). This has significantly increased the visibility of Ethiopia as a major regional economic actor in the East and Horn of Africa regions. Ethiopia's decision to move ahead with the GERD in 2011 was also partly an outcome of these structural changes that have taken place over the past decade.

By contrast, Egypt, once the most politically stable country in the basin and with by far the strongest economy, has experienced a different trajectory under which its political and economic influence has waned. The Egyptian revolution of 2011 and the subsequent years of political upheaval have led to a decline in the national economy, generalised social and political instability, changes in key Nile decision-making bodies and a more opaque regional strategy. For example, since 2011, the Egyptian Minister of Water Resources and Irrigation has changed five times, and according to several interviews conducted by the authors in Cairo (in 2011 and 2012), there has been an increasing involvement of foreign affairs and other high-level political circles in the Nile decision-making processes (see Tawfik 2016). This is thought to have contributed to an increased politicisation and securitisation of water issues. Although formal national water policies have changed little, attitudes towards neighbouring riparians and the cooperation process more generally have changed significantly, including returning to a more confrontational tone under President Muhammad Morsi's short presidency, when even the threat of military action against Ethiopia was expressed in public fora (BBC News 2013a; Al-Jazeera 2013). Taken together – Ethiopia's decision on GERD, Burundi's signature of the CFA in 2011, Sudan's new plans to increase irrigated agriculture, the support of South Sudan for the CFA, and CFA ratification by three upstream countries – these events represent a significant setback in Egypt's capacity to influence its neighbours (Cascão and Nicol 2013). At the same time, the somewhat equivocal reaction to the GERD project by Egypt – first an official rejection of the project, and then threats of sabotaging the dam, followed later by participation in trilateral negotiations and agreements – indicate that Egypt was initially unprepared to deal with such an emblematic event, particularly when distracted by intense internal political and economic change.

The third Eastern Nile riparian – Sudan – has perhaps experienced the largest economic and political changes of all, not least because of its subdivision, in 2011, into two independent states. From 2002 to 2005, the government of Sudan and the breakaway South Sudan Peoples' Liberation Movement negotiated a Comprehensive Peace Agreement (CPA), under which the Republic of South Sudan was born and further large-scale investment in the oil sector agreed. Massive oil revenues had hitherto allowed Khartoum to invest in large infrastructure in the country, during the early 2000s, including the Merowe hydropower dam, which was the first in a series of major planned hydraulic infrastructure projects (Verhoeven 2011). Taken together, the anticipated independence of South Sudan in 2011, expected reductions in oil revenues and the global food crisis of 2008 contributed to a renewed focus on agriculture in Sudan as a driver of development and encouraged more inward investment to its vast irrigation schemes, particularly on the Blue Nile. This could have huge implications for demand for water and for potential flows downstream to Egypt (McCartney *et al.* 2012). In order to proceed with these new plans, the Sudanese government leased several thousand hectares of land to private investors, embarked on heightening of the Roseires Dam to increase water storage capacity, and more recently, supported the GERD project, expecting that it would provide additional water for its ambitious new irrigation plans. In parallel (and partly as a consequence), political-diplomatic alliances with Egypt deteriorated, while those with Ethiopia improved, including several agreements on bilateral trade and economic integration. The internal changes in Sudan are of great hydropolitical importance and their magnitude has yet to be fully understood – as Sudan is becoming the 'kingmaker' in the Nile hydropolitical game, while negotiating water and hydraulic infrastructure to expand its large-scale national irrigation potential (Cascão and Nicol 2016).

Discussion on the GERD as an outcome of change

One of the main arguments of this article is that the GERD project is an outcome of multiple changes that were ongoing in the region before 2011. One can argue that economic and political changes at a national level in Ethiopia were key driving factors, and significantly contributed to a sense of urgency surrounding water resources development and hydraulic infrastructure in order to respond, in particular, to internal energy demands. In this section, however, we are more concerned with understanding and analysing how the pre-2011 'cooperation norms and processes' actually contributed to GERD becoming a reality.

GERD: an outcome of failed expectations?

It was mentioned in the first section that the cooperation process was progressing relatively well along both technical and political tracks until mid-2010. Under the NBI and its centres, the countries have adopted new joint and cooperative norms in the planning, management and development of its shared resources. The NBI had formed an all-inclusive cooperative platform where riparians could work together on ambitious projects, including several investment projects that could bring future socio-economic benefits (NBI 2011b). Despite the limited tangible results achieved up until then, the expectation remained that, in the medium or long term, benefits would be delivered through actual investments in hydraulic infrastructure (Earle *et al.* 2013; NBI 2012; Cascão 2012). But how much longer could the Nile riparians (in particular Ethiopia) wait for these investments to take shape in the face of growing national demands for economic development and pressure to develop the Nile water resources? It is useful to recall that when Ethiopia joined earlier multilateral cooperative processes in the Nile Basin, this was based on a strong belief that the political track would lead to a new comprehensive legal agreement for the basin and that the NBI would facilitate actual development of joint multipurpose infrastructure in the Ethiopian highlands (Amer 1997; Arsano and Tamrat 2005). In this context, Ethiopia's expectations that NBI/ENSAP investment would take place were high, and the government considered this to be a golden opportunity to develop its portion of the Blue Nile Basin. The hydraulic ambitions of Ethiopia can be traced back to the 1960s, when the US Bureau of Reclamation developed an in-depth study on the Ethiopian Blue Nile and identified major sites for hydropower and irrigation development (Waterbury 2002). In the 1980s and the 1990s, successive Ethiopian governments developed master plans that included a cascade of dams on the Blue Nile, namely, the Karadobi, Mabil, Mandaya and Border (the last being the most downstream and only 40 km from Sudan). Further in-depth studies confirmed the huge hydropower potential of the dams in this sub-basin (Block *et al.* 2007). However, for decades, Ethiopia was unable to muster sufficient economic, political and financial power to develop these projects (Cascão 2009).

Therefore, for Ethiopia, the ENSAP JMP was the first real opportunity to develop major hydropower on the Blue Nile, with the added advantage of doing it jointly with neighbouring riparians and in a spirit of collaboration and cooperation, while having access to external funding. As mentioned earlier, the 2008 JMP scoping study identified the cascade of hydropower dams on the Ethiopian Blue Nile as the projects that would generate the greatest opportunities and rewards for all the Eastern Nile countries (Blackmore and Whittington 2008). However, after 2008, the project reached stalemate and the ensuing hydropolitical 'crisis' of 2010 had immediate impacts on Ethiopia's expectations – including the curtailing of NBI activities and projects, and decreasing financial support for the overall cooperation process (Earle *et al.* 2013; NBI 2014). As a result, the possibility of moving ahead in the short term with NBI supported large-scale projects rapidly faded. In this context, implementation of the JMP and plans to jointly develop dams in the Ethiopian Blue Nile ground to a halt. According to interviews conducted by the authors with Ethiopian decision makers at the Ministry of Water Resources in Addis Ababa (in 2009 and 2010), the JMP project was losing political momentum, and Ethiopia understood that the possibility of developing the Blue Nile cascade through NBI/ENTRO was fast becoming more remote and unlikely. Accordingly, Ethiopian officials decided to go back to their pipeline of national projects for the Blue Nile, arguing that they were not in a position to wait any longer to develop the country's hydropower potential, taking account of energy demand in the fast-growing economy (Horn Affairs 2011). Consequently, at the end of 2010, the Ethiopia government made the decision to develop one of the dams, in approximately the same location as the Border Dam (identified in the 1964 USBR study, included in the 2008 JMP Scoping Study, but with different characteristics and dimensions, namely, its storage capacity), which was followed by a public announcement in April 2011.

In brief, Ethiopia's decision to move ahead in 2011 with the GERD as a national project can be considered as an outcome of failed expectations under the two parallel cooperation tracks. A comprehensive legal agreement was not endorsed, even after fourteen years of negotiations and NBI-identified investment projects had not materialised not least because of the uncertain institutional and legal situation after 2010. However, Ethiopia's decision to build the GERD as a national project did not preclude continued engagement in multilateral cooperation as the way forward to develop Nile waters. In fact, Ethiopia has continued without interruption to provide political, technical and financial support to the NBI and ENTRO, while reiterating the need for a comprehensive framework agreement to support equitable and reasonable utilisation of the Nile water resources (Ethiopian Herald 2015a; Horn Affairs 2013).

GERD: an outcome of changes in the regional power balance?

Power asymmetry, and changes in these asymmetries, can affect actual interactions between riparian states in a transboundary river basin (Zeitoun and Mirumachi 2008; Zeitoun et al. 2014). A core argument of this article is that cooperation processes initiated in the 1990s in the Nile Basin eventually contributed to reducing existing asymmetries and changed the de facto balance of power between upstream and downstream neighbours. Two decades of multilateral cooperation contributed to strengthening upstream riparian capacities in many ways, including bolstering their ideational and bargaining powers, including the power to influence the knowledge agenda, discourse and negotiations at a basin level (Zeitoun et al. 2011; Cascão 2009). In terms of ideational power, at the beginning of the 1990s, asymmetries were great. Both Egypt and Sudan had an in-depth and long-standing technical establishment that had built up substantial knowledge of the Nile's water and land resources, including knowledge of current utilisation, future trajectories and scenarios, and potential for development. At the same time, knowledge upstream was more limited and less technically sophisticated – hindering capacity to identify and study key potential development projects. However, from the early 1990s onwards, countries including Ethiopia began substantial efforts to improve national capacities (Arsano 2007; Waterbury 2002).

Fifty years down the line, and after many NBI projects including in-depth studies focusing on optimisation of resources and the role of future infrastructure in that optimisation, upstream riparian states are now more empowered as 'knowledge influencers' within wider transboundary cooperation. This includes being more equipped with enhanced analytical tools, and having substantially more institutional capacity for communication, information, management and analysis (NBI 2014; Qaddumi 2008). Interviews conducted in several forums by the authors with representatives from upstream countries indicate that the NBI cooperation (through its numerous knowledge-oriented programmes and tools) contributed in significant ways to reducing past knowledge gaps, and ultimately has rendered upstream riparians more empowered to influence dialogue, set agendas and undertake regional policy-making.

However, it is in terms of bargaining power, that is, power to influence the negotiation process and outcomes (Zeitoun *et al.* 2011, 2017), that these changes are most visible. The capacity of upstream countries to influence transboundary negotiations and their outcomes differs greatly from twenty years ago. On the one hand, the NBI and CFA processes have opened up the possibility of bringing issues onto the agenda under different negotiation processes that, hitherto, were thought not possible. These include issues of hydraulic development upstream and the long-avoided issue of a new legal framework agreement based on the principle of equitable utilisation of resources (Cascão and Zeitoun 2010b). Until the mid-1990s, a debate on these two topics would invariably encounter strong resistance from downstream riparians. On the other hand, and less anticipated, the two cooperation tracks have also opened up the opportunity for a more complex set of 'sub'-alliances between upstream countries, and even an eventual alliance between Sudan and Ethiopia. An example of the first is the coalition of interests that brought together all the upstream riparians in 2010 to sign the CFA. The other example is the growing alliance between Sudan and Ethiopia, based on increased understanding of the substantial benefits that Sudan can reap in technical and economic terms from bilateral cooperation with Ethiopia. This was partially made possible because the NBI and ENTRO platforms allowed increased dialogue between the two neighbours and NBI-driven studies contributed to generating scientific evidence on the multiple benefits to be accrued to Sudan from cooperative investments (Hamad and El-Battahani 2005; NBI 2011a).

Concerning decreasing power asymmetries, in the case of Ethiopia this is much more visible in political and diplomatic actions that the country has taken in the past decade. It was Ethiopia that pushed hard for the CFA to be signed by almost all upstream countries and played a strong diplomatic role in bringing this new type of alliance to the fore. Ethiopia has also aligned with the NELSAP riparian countries and proceeded with CFA signature in spite of reservations expressed by its two downstream neighbours. The official discourse of Ethiopia has hitherto been one of assuming leadership of the 'upstream bloc', namely, having the prime minister declare that: 'the upper riparian countries that signed the CFA are highly desirous for the Agreement to be ratified and implemented [and] Ethiopia has to be exemplary to other riparian countries by ratifying the agreement' (Horn Affairs 2013). Accordingly, in June 2013, Ethiopia was the first country to ratify the CFA. These are examples of how Ethiopia's bargaining powers have increased dramatically over the past decade (Zeitoun *et al.* 2014). But in discursive and ideational terms, the cooperation process has also provided Ethiopia with an opportunity to sit at several negotiating tables. Without renouncing the supremacy of the CFA as a mechanism for effective cooperation in the Nile Basin, simultaneously Ethiopia has leveraged its diplomatic strengths under the GERD process and been able to transform an initially unilateral project into a tripartite process with Sudan and Egypt (as analysed in next section). In both its multilateral and its bilateral/trilateral relations, it is possible to conclude that Ethiopia has substantially developed its capacity to influence the current state of affairs in the Nile Basin, and has done so through foregrounding two issues: the need for a new legal framework in the basin and the benefits of water infrastructure development upstream for all riparians.

In conclusion, the authors of this article consider that the two tracks to multilateral cooperation (NBI and CFA) have partially contributed to changing relations between riparians through increasing upstream riparians' knowledge of resources at their disposal and potential development opportunities, and at the same time, increasing their overall capacity to shift perceptions, wider discourses and the broader agenda at a basin-wide level. This has reduced the asymmetric balance of bargaining power between riparians and increased opportunities for alliance building at bilateral and multilateral levels. The twenty years of cooperation processes on the Nile since the mid-1990s have been a partial – but significant – contributory factor in Ethiopia's capacity to influence the hydro-political landscape of the Eastern Nile, of which the GERD is the clearest and most immediate manifestation.

GERD-related cooperation norms and processes

In April 2011, Ethiopia's prime minister publicly announced his country's decision to build the Grand Ethiopian Renaissance Dam. Although project identification and planning had been carried out by the Ethiopian government since 2010 and in line with the country's five-year Growth and Transformation Plan (GTP) launched in the same year (FDRE 2010; Matthews *et al.* 2013), the technical details of this 'new' dam and its possible hydrological impacts were still largely unknown. Nevertheless, questions surrounding the short- and long-term political impacts were already being posed: Would this new project – the first large-scale dam in the Ethiopian portion of the Blue Nile – lead the Eastern Nile countries back to patterns of hydropolitical dispute that had been the norm before the 1990s? Would this mean the end of multilateral cooperation between the Nile countries? At first there were concerns that Ethiopia and downstream countries could enter into serious conflict, taking account of the dam's size and the fact that it would be located in the strategic Blue Nile Basin. Despite the moderate tone of Ethiopia's prime minister in the 2011 politically crafted speech alluding to the win–win benefits of the dam project for all three countries and inviting the two neighbours to join and co-finance the GERD project (Zenawi 2011), it soon became clear that Ethiopia would implement the project with or without downstream financial and political support. Yet, as soon as May 2011, the three neighbours had decided to initiate a trilateral process to deal with the GERD project – establishing a technical dialogue that would later evolve into a negotiated political process. The first part of this section looks at the original and unique features of the new trilateral approach in terms of cooperation norms and processes. The last section discusses the impacts that the GERD could have on multilateral institutions (such as the NBI) and the wider process of cooperation.

Trilateral cooperation: from technical talks to political negotiations

A superficial analysis of the media discourse on GERD during 2011 would suggest that the Eastern Nile Basin was experiencing a new hydropolitical conflict. The fact is that behind the scenes the three riparians had also begun a process of trilateral collaboration, with new rules of the Nile game. The dialogue began soon after the announcement of the GERD and was followed by a visit in May 2011 by the interim Egyptian Prime Minister, Essam Sharaf, to Addis Ababa. This culminated in a joint statement that the Eastern Nile neighbours were willing to establish a trilateral process to analyse the impacts of the dam (Al-Ahram Weekly 2011; Ahram Online 2011). At the end of May 2011, Ethiopia and its two neighbours agreed to initiate trilateral talks on technical issues regarding the dam, and in October the countries agreed to set up a trilateral technical committee to assess the possible impacts of the dam, exchange technical expertise, foster cooperation and ultimately to boost regional development (Sudan Tribune 2011). Soon after, the three ministers of water resources agreed on the terms of reference and rules of procedures for the establishment of an international panel of experts (IPoE), and within two months the governments had each nominated two national experts and selected four international experts. The IPoE held its first official meeting in January 2012. As a point of departure, this was a positive achievement: the countries were not just willing but also able to agree on a mechanism to conduct a joint assessment of the impacts of the new infrastructure. Between May 2012 and May 2013, the IPoE would hold six regular meetings and four visits to the dam site; the final report was released on 31 May 2013. The report covered several technical issues, including downstream concerns over dam safety, and recommended two additional studies: assessment of transboundary environmental and socio-economic impacts; and a new hydrological model study (IPoE 2013; also see MIT 2014; Wheeler *et al.* 2016; and Zhang *et al.* 2016).

The reactions of Sudan and Egypt to the IPoE report were nevertheless divergent. In Sudan, it was followed by numerous official declarations of support by the government, downplaying the negative impacts and praising the benefits of the GERD for Sudan, namely, its potential to regulate flows and contribute to expanding irrigated agriculture along the Sudanese Blue Nile (Sudan Tribune 2013a, 2013b). In Egypt, where a new government led by President Morsi was already in place and held different views on the GERD from previous interim governments, the reaction to the IPoE report was marked by criticism. As in the case of the earlier JMP study, Egypt similarly disputed the technical validity of the joint study (though Egyptian experts had formed part of the team) and decided to conduct alternative studies, which came to different conclusions from the joint study and highlighted the potential negative hydrological impacts of the GERD (MFA Egypt 2014). In brief, technical talks and joint mechanisms were in place when the GERD process reverted to issues of securitisation and older discourses on 'water wars' emerged (BBC News 2013a, 2013b). The elected government of President Morsi was not keen to accept the GERD as a *fait accompli*, and new political and even veiled military threats were made against Ethiopia and the dam, rapidly souring bilateral diplomatic relations between the two countries. This was the moment that Sudan, the midstream riparian expressed strong belief in the benefits the GERD, and started adopting the unofficial role as mediator, offering its 'good offices' to bring the parties together to agree on how to resume trilateral talks and adopt the IPoE's recommendations (Al-Ahram Weekly 2014; Sudan Tribune 2014). Trilateral meetings at the end of 2013 and again in January 2014 were postponed or failed to reach any conclusion. These were only resumed when a new government was elected in Cairo (in May 2014) and the newly installed President El-Sisi agreed that it was time to reactivate the GERD tripartite meetings. However, this time, diplomats would be fully involved and the trilateral meetings would not only be technical but also of a political and legal nature – becoming, in effect, a 'six-party' meeting involving the ministries of water resources and foreign affairs of the three countries. In August 2014, the three countries resumed the trilateral meetings and subsequently met on six occasions, until the end of December 2015 (Salman 2016).

The main conclusion to draw from this trajectory of change is that despite a level of openness from Egypt to take part in technical talks soon after the GERD was announced in 2011 (establishing a joint team to study the impacts of the dam), not long after technical cooperation was once again overridden by political considerations, leading to securitisation of the trilateral cooperation process and its transformation into a complex technical-cum-political process. Nevertheless, the trilateral process did lead to an agreement being signed by the three riparian states, as discussed next.

An agreed trilateral framework and new cooperation norms

The trajectory to trilateral interaction between 2011 and 2014 reveals the coexistence of cooperation and conflict, which is very often the case in transboundary water interaction (Zeitoun and Mirumachi 2008). In the Eastern Nile Basin there is on the one hand, a form of all-inclusive technical and political cooperation, but on the other hand, lingering diplomatic conflicts. As in the past, the countries were operating without a common legal agreement in place. The trilateral cooperation initiated in May 2011 can be considered an *ad hoc* type of cooperation and largely project-specific in nature. This takes us back to the importance of legal agreements in the transboundary hydropolitical interaction between Nile riparians. The official Ethiopian position is that the CFA remains the legal framework of reference and the only one that institutionalises all guiding legal principles on water resources management and development in the Nile Basin (Ethiopian Herald 2015b). However, this is not the case for Egypt and Sudan, both of which have not signed the CFA document and still have strong reservations regarding parts of the text. Egypt maintains a strong stance against the CFA and is still calling on upstream neighbours to renegotiate

(Ahram Online 2015). Generally speaking, Egypt remains in favour of agreements that offer guarantees that developments upstream will not affect what it calls 'national water security'. These two very different stances, however, did not prevent high-level officials of the three countries from signing a new legal instrument, which indicates the different positions are not completely irreconcilable.

Several rounds of high-level talks at the beginning of 2015 culminated in the preliminary draft of an agreement for the GERD, and in early March, the ministers of water resources and foreign affairs jointly announced that a deal was being finalised, representing the beginning of a new era of cooperation. On 23 March 2015, the Declaration of Principles (DoP) for the GERD was signed in Khartoum by the three heads of state. This was considered an historic deal, bringing together for the first time the three Eastern Nile countries around guiding principles on cooperative relations. Among its ten principles, several are commonly accepted principles of international water law, such as 'no significant harm' and 'equitable and reasonable utilization'; but they also include principles more related to technical issues such as dam security, dam filling, operations policy and exchange of information (Agreement 2015). To reaffirm these principles, overcome challenges and give political drive to the DoP, both water and foreign affairs ministers organised new rounds of trilateral meetings in Khartoum in late 2015. Finally, on 28 December 2015, the Khartoum Accord was formally signed (Sudan Tribune 2015; Salman 2016). In hydropolitical terms, these two legal instruments cement a new hydropolitical reality in the Eastern Nile, one that includes the GERD as a fact on the ground, recognised by all three Eastern Nile riparians. This is a very significant 'tipping point' in that it establishes a very different set of norms and processes than had hitherto existed under the NBI and ENTRO institutions. Technical cooperation in this manner is more specific and project-oriented than the studies conducted under ENTRO, for example. Cooperation on GERD is country-driven and has almost no involvement from external partners, counting only on technical and financial support from the countries themselves. Egypt has in effect returned to trilateral collaboration and talks with the neighbours and has shown openness to concrete discussion of upstream hydraulic projects. Moreover, the three countries have successfully reached an agreement at the highest political level, the ultimate outcome of which is a new large-scale dam on the Blue Nile that started life as an idea at a multilateral level under the JMP/ENTRO project, before re-emerging as a national-level Ethiopian project, and was later enshrined in a trilateral legal-political agreement.

In conclusion, the authors of this article consider that the in-depth analysis of the process regarding the GERD provides evidence that different cooperation norms have been adopted and have subsequently brought about key changes in transboundary interaction in the (Eastern) Nile Basin. Ultimately, it has contributed to reactivating trilateral relations between Egypt, Ethiopia and Sudan that had been in stalemate since Egypt and Sudan 'froze' participation in the NBI in mid-2010.

GERD: a shaper of future cooperation developments

This final, more forward-looking section explores how near completion of the GERD and emerging relationships between Ethiopia, Egypt and Sudan are likely to shape the future of transboundary cooperation in the Nile Basin. This includes increasing understanding of the tangible economic benefits that multi-lateral cooperation can offer to regional and national economies and by raising awareness among all riparians of the need for comprehensive and inclusive agreement on an institution that can guide future basin-level developments.

As an established part of the landscape in Ethiopia's Blue Nile, the sheer physical scale and the policy impact of the GERD are unparalleled since the construction of the High Aswan Dam in Egypt over fifty years ago. Although it will take many years for the dam's full impact (and wider implications) to be felt and understood across the three Eastern Nile countries – as did the High Aswan Dam's – it is nonetheless useful to sketch out potential scenarios for cooperation. First, economically speaking, the dam will have a major impact on the availability of energy in the Eastern Nile Basin and even beyond. The GERD will substantially increase available hydropower, with immediate benefits for Sudan (the grid to which is already connected with Ethiopia) and, potentially, longer-term benefits for Egypt, if it connects to the existing regional power grids, including those being planned under the Eastern Africa Power Pool. Therefore, it is possible that the GERD (and other future hydropower dams upstream) could provide substantial incentives for future economic cooperation and trade between Ethiopia, Sudan, Egypt and other Nile riparian countries. In the longer term, the unlocking of economic development potential in Ethiopia and Sudan through energy production and trade could support wider structural changes in both economies, providing new trade and investment opportunities for Egypt. In common with other upstream countries, Ethiopia is currently experiencing a transition from a largely agrarian and commodity-dominated economy with limited employment generation to a more diversified, urbanised and industrial-ised economy that can offer several opportunities for its rapidly growing and economically active population. Foreign direct investment from neighbouring countries, including Egypt, could contribute to this goal and provide a basis for the deepening of regional economic and political integration, eventually providing net benefits for all regional economies. GERD-related energy production and trade could then be considered as the initial trigger of this wider process.

Besides, the construction and operation of the GERD itself is likely to demonstrate a set of wider potential economic benefits that other similar projects could replicate. This would further strengthen the case for more pooling of power resources, joint development of multipurpose projects and the co-management of other basin resources (including land for irrigation and wider ecosystems). The emphasis on coordination and cooperation to generate and trade in economic goods is an essential underpinning to future basin-wide cooperation processes. First, the GERD and its anticipated outcomes in terms of energy, economic development and regional trade will help demonstrate in the most tangible way that joint developments can bring shared benefits – in a way that earlier 'fast-track' projects under the NBI/ENTRO managed only to partially deliver, due to their smaller scale and slower implementation. Second, at a political level, the diplomatic process triggered by the GERD is likely to have a number of long-term consequences for cooperation. Under the DoP signed in March 2015, Egypt has, in effect, decided to enter into a formal legal political arrangement with its upstream neighbour Ethiopia which, *de jure*, lies outside the boundaries of the 1959 bilateral agreement with Sudan and therefore seems to suggest a willingness to negotiate future decisions in legally binding agreements that go beyond existing treaties. For upstream riparians, this could signal a new Egyptian accommodation with future infrastructure developments, enshrining the realpolitik of recent years in new rules of the game – Egypt seemingly recognising the inevitability of upstream developments and around which it has started negotiating separate agreements. The GERD process has, therefore, contributed to extending the discussion about tangible and future hydraulic developments in the Basin at a broader level. At the same time, the GERD has contributed to countries sharing more publicly their water and economic needs and national plans (including new directions from Sudan on future water and development priorities). Overall, this is contributing to a more realistic debate on the shape and direction of future joint development of the Nile water resources and enabling a shift from cooperation principles to development practice.

From an institutional perspective, the GERD has also helped champion dialogue over disputes. It has shown that countries that have held largely antagonistic positions can, through dialogue, achieve compromise and agreement. This is an extremely important departure from hitherto prevailing situations in which countries were reluctant to commit to negotiated agreements. In the longer term, this might contribute to strengthening the NBI's *raison d'être*, illustrating a tilt towards cooperation rather than conflict in spite of convoluted past hydropolitical relations. The NBI can, therefore, legitimately claim to be the most significant, durable and truly basin-wide institution in which to enshrine this cooperation. For example, Sudan's response to GERD has strengthened rather than weakened the NBI, not least because it illustrates the current relevance of the NBI over the more historically bound bilateral 1959 Agreement with Egypt. This institutionalisation of a 'new basin politics' suggests that the NBI will remain the logical powerhouse for future projects on the Nile Basin. One can conclude that the GERD process generated a new, more forthright, debate in the Eastern Nile region, and across the Nile Basin more generally. Above all, it shows that negotiation and compromise are possible, even in the most complex political situations.

There are also other important learning points from the GERD experience. It proves that countries can cooperate simultaneously under different platforms. This includes promoting dialogue on a specific project under one platform (as in the GERD case) while simultaneously developing projects under other multilateral arrangements (as Ethiopia and Sudan continue to do under ENTRO and the NBI, but also under other regional institutions, including the Eastern Africa Power Pool and Intergovernmental Authority on Development). However, the GERD process also reminds the wider global community that tripartite negotiations surrounding the dam represent substantial transaction costs in the absence of prior agreements between countries. In future, the proliferation of such bilateral or trilateral negotiations and agreements specific for each planned project would represent huge (and avoidable) transaction costs, which pre-existing multilateral arrangements could help to minimise or prevent. Building into the NBI processes forward planning, assessment and evaluation of future projects not only supports more coherent policy and planning at the whole basin level, but also paves the way for political agreement across a range of other projects and avoids more piecemeal negotiations over separate activities and projects.

Regarding Egypt, the greatest consumer of and most dependent country on Nile waters, there is every sign that future upstream developments will now take place, regardless of expressed opposition downstream. The GERD is symbolic of a strategic power shift within the basin towards upstream states that had been ongoing for many years, and under which greater demands for energy resources and the achievement of food security increasingly drive economic and political decision-making in conjunction with newly acquired financing and implementation capacity, namely, to build large-scale infrastructure. This reduces the hitherto political leverage capacity of downstream states over conventional financing partners and institutions. Achieving an accommodation with these new realities is now a key challenge for Egypt. This includes avoiding a patchwork of individual agreements under each future project upstream, which might otherwise generate high transaction costs for all parties concerned and require substantial time and effort to achieve.

For wider Nile transboundary cooperation, inadvertently the GERD process provides a strategic opportunity. On many levels – including economic, political and diplomatic – there is a more pressing need now for a transboundary water regime informed by a basin-wide approach and enshrined in a transboundary agreement housed within a permanent river basin organisation. After sixteen years of experience, the NBI has strong foundations on which to build such an approach, including its own strategic assessments and the identified opportunities for joint investments. Though not without controversy, the GERD process has, perhaps inadvertently, shone a light back onto the NBI and the need to continue to unlock the potential for true basin-wide cooperation.

Disclaimer

The views expressed in this article are those of the authors and do not necessarily reflect the official policy or position of the respective institutions.

References

AfDB, OECD and UNDP (2015). Ethiopia 2015. In AfDB, OECD and UNDP (eds), *African Economic Outlook 2015: Regional Development and Spatial Inclusion*. Paris: OECD.

AFP (2014, 19 June). Sudan asks Egypt to Rejoin Nile Basin Body. *Ahram Online*.

Agreement (1959). *Agreement between the Republic of the Sudan and the United Arab Republic for the Full Utilization of the Nile Waters*. Signed at Cairo, Egypt, 8 November.

Agreement (2010). *Cooperative Framework Agreement on the Nile River Basin Cooperative Framework*. Signed at Entebbe, Uganda, 14 May.

Agreement (2015). *Agreement on Declaration of Principles between the Arab Republic of Egypt, the Federal Democratic Republic of Ethiopia and the Republic of the Sudan on the Grand Ethiopian Renaissance Dam Project*. Signed at Khartoum, Sudan, 24 March.

Ahram Online (2011, 3 May). Egypt's Water Minister Hails New Era in Egyptian–Ethiopian Relations.

Ahram Online (2015, 18 December). Egypt Still Rejects Nile Basin's Cooperative Framework Agreement: Irrigation Minister.

Al-Ahram Weekly (2011, 19 May). Nile Row Easing.

Al-Ahram Weekly (2014, 19 February). Sudanese FM says his Country Impartial in Egypt–Ethiopia Crisis over Renaissance Dam.

Al-Ahram Weekly (2016, 28 January). Ethiopia's Game Plan on the Dam.

Al-Jazeera (2013, 11 June). Egypt warns Ethiopia over Nile Dam.

Amer, S.E. (1997). Cooperation in the Nile Basin: Appropriate Legal And Institutional Framework. In *Proceedings of the 5th Nile 2002 Conferences*, Addis Ababa 24–28 February 1997, 325–336.

Arsano, Y. (2007). *Ethiopia and the Nile: Dilemmas of National and Regional Hydropolitics*. Zurich: Swiss Federal Institute of Technology.

Arsano, Y. and Tamrat, I. (2005). Ethiopia and the Eastern Nile Basin. *Aquatic Sciences*, 67(1), 15–27.

Bayeh, E. (2015). New Development in the Ethio-Egypt Relations over the Hydro-Politics of Nile: Questioning its True Prospects. *International Journal of Political Science and Development*, 3(3), 159–165.

BBC News (2013a, 4 June). Egyptian Politicians Caught in On-Air Ethiopia Dam Gaffe.

BBC News (2013b, 10 June). Egyptian Warning over Ethiopia Nile Dam.

Blackmore, D. and Whittington, D. (2008). *Opportunities For Cooperative Water Resources Development on the Eastern Nile: Risks And Rewards. An Independent Report of the Scoping Study Team to the Eastern Nile Council of Ministers*. Washington, DC: World Bank.

Block, P., Strzepek, K. and Rajagopalan, B. (2007) *Integrated Management of the Blue Nile Basin in Ethiopia: Hydropower and Irrigation Modeling*. IFPRI Discussion Paper 700. Washington, DC: International Food Policy Research Institute.

Brunnée, J. and Toope, S.J. (2002). The Changing Nile Basin Regime: Does Law Matter?. *Harvard International Law Journal*, 43(1), 105–159.

Cascão, A.E. (2009). Changing Power Relations in the Nile River Basin: Unilateralism vs. Cooperation?. *Water Alternatives*, 2(2), 245–268.

Cascão, A.E. (2012). Nile Water Governance. In S.B. Awulachew, V. Smakhtin, D. Molden and D. Peden (eds), *The Nile River Basin: Water, Agriculture, Governance and Livelihoods*. Abingdon: Routledge-Earthscan.

Cascão, A.E. and Nicol, A. (2013). *The Political Context for Cooperation in the Nile Basin: A State of Flux, a New Dynamism*. Unpublished Report.

Cascão, A.E. and Nicol, A. (2016). Sudan, 'Kingmaker' in a New Nile Hydropolitics: Negotiating Water and Hydraulic Infrastructure to Expand Large-Scale Irrigation. In E. Sandström, A. Jägerskog and T. Oestigaard (eds), *Land and Hydropolitics in the Nile River Basin: Challenges and New Investments*. Abingdon: Routledge-Earthscan.

Cascão, A.E. and Zeitoun, M. (2010a). Power, Hegemony and Critical Hydropolitics in the Middle East and North of Africa Region. In A. Earle, A. Jägerskog and J. Öjendal (eds), *Transboundary Water Management: Principles and Practice*. Abingdon: Routledge-Earthscan.

Cascão, A.E and Zeitoun, M. (2010b). Changing Nature of Bargaining Power in the Hydropolitical Relations in the Nile River Basin. In A. Earle, A. Jägerskog and J. Öjendal (eds), *Transboundary Water Management: Principles and Practice*. Abingdon: Routledge-Earthscan.

Earle, A., Nordin, K., Cascão, A.E., Rukundo, D., Seide, W. and Björklund, G. (2013). *Independent Evaluation of the Nile Basin Trust Fund: Final Report*. Washington, DC: World Bank Group.

The East African (2009, 22 June). Egypt, Sudan Renege on New Nile Pact.

Eldaw, A. and Fekade, W. (2009). *Sustainable Transboundary Basin Development as a Strategy for Climate Change-Induced Conflict Prevention: Reflections from Eastern Nile*. Paper presented in Workshop on Climate Change and Transboundary Water Resource Conflicts in Africa, Mombasa, Kenya.

ENSAP (2007). *A Call for Accelerating Action on Eastern Nile Cooperation: A Report on the First JMP Regional Parliamentarian Exchange Visit*. Addis Ababa: ENTRO.

ENSAP (2009). *Factsheets of the 8 IDEN Projects*. Addis Ababa: ENTRO.

Ethiopian Herald (2015a, 2 March). Ethiopia: Shared Water Resource Optimal Benefit can be Reaped Only When the Countries Genuinely Cooperate.

Ethiopian Herald (2015b, 19 March). Ethiopia: GERD is the Symbol of Ethiopia's Commitment to Mutual Development of the Nile.

Federal Democratic Republic of Ethiopia (FDRE) (2010). *Growth and Transformation Plan (GTP) 2010/2011–2014/2015*, Ministry of Finance and Economic Development, Addis Ababa, Ethiopia.

Financial Times (2015, 26 April). Ethiopia Predicts Record $1.5bn Overseas Direct Investment in 2015.

Gebreluel, G. (2014). Ethiopia's Grand Renaissance Dam: Ending Africa's Oldest Geopolitical Rivalry?. *The Washington Quarterly*, 37(2), 25–37.

GTZ (2007). *Donor Activity in Transboundary Water Cooperation in Africa: Results of a G8-Initiated Survey 2004–2007*. Report Commissioned by the German Federal Ministry for Economic Cooperation and Development, GTZ, Eschborn, Germany.

Hamad, O.E. and El-Battahani, A. (2005). Sudan and the Nile Basin. *Aquatic Sciences*, 67(1), 28–41.

Horn Affairs (2011, 2 April). Ethiopia's Dam on Nile Launched and Text of Meles Zenawi Speech.

Horn Affairs (2013, 21 April). Nile: Ethiopia Pokes Egypt Taking the Last Step to Ratify CFA.

International Panel of Experts (2013). *Final Report on the Grand Ethiopian Renaissance Dam*. Addis Ababa, Ethiopia.

McCartney, M., Alemayehu, T., Easton, Z. and Awulachew, S. (2012). Simulating Current and Future Water Resources Development in the Blue Nile River Basin. In S.B. Awulachew, V. Smakhtin, D. Molden and D. Peden (eds), *The Nile River Basin – Water, Agriculture, Governance and Livelihoods*. Abingdon: Routledge-Earthscan.

Makonnen, T. and Lulie, H. (2014). *Ethiopia, Regional Integration and the COMESA Free Trade Area*. SAIIA Occasional Paper 198. Johannesburg: South African Institute of International Affairs.

Matthews, N., Nicol, A. and Seide, W. (2013). Constructing a New Water Future?: An Analysis of Ethiopia's Current Hydropower Development. In J.A. Allan, M. Keulertz, S. Sojamo and J. Warner (eds), *Handbook of Land and Water Grabs in Africa*. London: Routledge.

Ministry of Foreign Affairs (MFA) of Egypt (2014). *Egypt's Perspective Towards the Ethiopian Grand Renaissance Dam Project (GERDP)*. Ministry of Foreign Affairs, Cairo, Egypt.

MIT (2014). *Grand Ethiopian Renaissance Dam: An Opportunity for Collaboration and Shared Benefits of the Eastern Nile Basin. An Amicus Brief to the Riparian Nations of Ethiopia, Sudan and Egypt from the International, Non-Partisan Eastern Nile Working Group.* 13–14 November 2014. MIT and Abdul Lateef Jameel World Water and Food Security Lab, Cambridge, MA.

Nicol, A. and Cascão, A.E. (2011). Against the Flow: New Power Dynamics and Upstream Mobilisation in the Nile Basin. *Review of African Political Economy*, 38(128), 317–325.

Nicol, A., van Steenbergen, F., Sunman, H., Turton, A.R., Slaymaker, T., Allan, J.A., de Graaf, M. and van Harten, M. (2001). *Transboundary Water Management as an International Public Good*. Stockholm: Ministry of Foreign Affairs.

Nile Basin Initiative (NBI) (1999). *Policy Guidelines for the Nile River Basin Strategic Action Program*. Prepared by the NBI secretariat in cooperation with the World Bank, NBI, Entebbe, Uganda.

NBI (2002). *Nile Basin Act*. Retrieved from http://faolex.fao.org/docs/pdf/uga80648.pdf.

NBI (2009). *Growing Cooperation Through Joint Actions – NBI Annual Report*. NBI, Entebbe, Uganda.

NBI (2011a). *Sudan, the NBI and the Benefits of Cooperation (NBI Country Papers)*. Entebbe, Uganda.

NBI (2011b). *Corporate Report*. Entebbe, Uganda.

NBI (2012). *NBI Overarching Strategic Plan 2012–2016*. Entebbe, Uganda.

NBI (2014). *Nile Cooperation: Opportunities and Challenges (NBI Flagship Paper)*. Entebbe, Uganda.

NBI (2016). Nile Cooperation: Gateway to Regional Integration. *NBI News*, 24 February.

Qaddumi, H. (2008). *Practical Approaches to Transboundary Water Benefit Sharing Working Paper 292*. London: Overseas Development Institute (ODI).

Reuters (2009, 28 July). Egypt says Historic Nile River Rights not Negotiable.

Salman, S.M.A. (2013). The Nile Basin Cooperative Framework Agreement: A Peacefully Unfolding African Spring?. *Water International*, 38(1), 17–29.

Salman, M.A.S. (2016). The Grand Ethiopian Renaissance Dam: The Road to the Declaration of Principles and the Khartoum Document. *Water International*, 41(4), 512–527.

State Information System (SIS), Egypt (2011, 3 March). Egypt and its Historical Rights in Nile Water.

State Information System (SIS), Egypt (2016, 24 February). Egypt: Irrigation Minister Highlights Egypt's Cooperation with Nile Basin Countries.

Sudan Tribune (2010, 27 June). Sudan Freezing its Membership in the Nile Basin Initiative.

Sudan Tribune (2011, 25 October). Sudan: Country Agrees to Tripartite Committee over Ethiopia's Nile Dam.

Sudan Tribune (2012a, 9 November). Agreement Reached on Eastern Nile Basin cooperation.

Sudan Tribune (2013a, 30 May). Sudan Downplays Negative Impact of Ethiopian Dam Project.

Sudan Tribune (2013b, 9 June). Sudan Reiterates Support of Ethiopian Dam Plans.

Sudan Tribune (2014, 24 August). Egypt wants Sudan to Mediate in Nile Water Tripartite Meeting.

Sudan Tribune (2015, 29 December). Sudan, Egypt and Ethiopia Reach Agreement on Renaissance Dam.

Tawfik, R. (2015). *Revisiting Hydro-Hegemony from a Benefit-Sharing Perspective: The Case of the Grand Ethiopian Renaissance Dam* (DIE Discussion Paper 5). Bonn: Deutsches Institut für Entwicklungspolitik (DIE).

Tawfik, R. (2016). The Grand Ethiopian Renaissance Dam: A Benefit-Sharing Project in the Eastern Nile?. *Water International*, 41(4), 574–592.

Verhoeven, H. (2011). *Black Gold for Blue Gold?: Sudan's Oil, Ethiopia's Water and Regional Integration*. London: Chatham House.

Waterbury, J. (2002). *The Nile Basin: National Determinants of Collective Action*. New Haven, CT: Yale University Press.

Wheeler, K.G., Basheer, M., Mekonnen, Z.T., Eltoum, S.O., Mersha, A., Abdo, G.M., Zagona, E.A., Hall, J.W. and Dadson, S.J. (2016) Cooperative Filling Approaches for the Grand Ethiopian Renaissance Dam. *Water International*, 41(4), 611–634.

Whittington, D., Waterbury, J. and Jeuland, M. (2014). The Grand Renaissance Dam and Prospects for Cooperation on the Eastern Nile. *Water Policy*, 16(4), 595–608.

World Bank. (2009). *Eastern Nile First Joint Multipurpose Program Identification (JMP1 ID) – Project Information Document*. Washington, DC: World Bank.

Zeitoun, M. and Mirumachi, M. (2008). Transboundary Water Interaction I: Reconsidering Conflict and Cooperation. *International Environmental Agreements*, 8(4), 297–316.

Zeitoun, M., Mirumachi, N. and Warner, J. (2011). Transboundary Water Interaction II: Soft Power Underlying Conflict And Cooperation. *International Environmental Agreements*, 11(2), 159–178.

Zeitoun, M., Warner, J., Mirumachi, N., Matthews, N., McLaughlin, K., Woodhouse, M., Cascão, A.E. and Allan, J.A. (2014). Transboundary Water Justice: A Combined Reading of Literature on Critical Transboundary Water Interaction and 'Justice', for Analysis and Diplomacy. *Water Policy*, 16(S2), 174–193.

Zeitoun, M., Cascão, A., Warner, J., Mirumachi, N., Matthews, N., Menga, F. and Farnum, R. (2017). Transboundary Water Interaction III: Contest and Compliance. *International Environmental Agreements: Politics, Law and Economics*, 17(2), 271–294.

Zenawi, M. (2011). Speech on Launching the GERD. Retrieved from www.meleszenawi.com/ethiopian-pm-meles-zenawi-speech-on-launching-gerd-text-and-videos/.

Zhang, Y., Erkyihun, S.T. and Block, P. (2016) Filling the GERD: Evaluating Hydroclimatic Variability and Impoundment Strategies for Blue Nile Riparian Countries. *Water International*, 41(4), 593–610.

6 GERD and hydropolitics in the Eastern Nile

From water-sharing to benefit-sharing?

Rawia Tawfik and Ines Dombrowsky

Introduction

In his speech on launching the Grand Ethiopian Renaissance Dam (GERD) in 2011, the late Ethiopian Prime Minister Meles Zenawi (2011) emphasised that the benefits that would accrue from the dam would not be restricted to Ethiopia, but would extend to all neighbouring states, particularly to the downstream riparians, Sudan and Egypt. He went further to argue that the GERD would resolve the differences over equitable utilisation of the Nile water resources since it would increase these resources by reducing evaporation and regulating the flow, and provide cheap energy to downstream countries (Zenawi 2011). His successor, Hailemariam Desalegn, reiterated the same message and even argued that Ethiopia considers the GERD to be a 'joint ownership' since its benefits extend to other Eastern Nile countries, Sudan and Egypt (Sudan News Agency 2013). Although the Ethiopian government had turned down an Egyptian proposal to participate in the financing and management of the dam on the grounds that the operation of the dam is a matter of sovereignty and that a joint administration is 'unacceptable' (Ministry of Foreign Affairs of the Arab Republic of Egypt 2014a; Horn Affairs 2014), it has continued to use the language of benefit-sharing to market the project regionally and internationally (Adhanom 2014).

It was not only the Ethiopian politicians who argued for sharing the benefits of the GERD. Scholars have long maintained that building dams on the Blue Nile would generate benefits to the river and to Eastern Nile countries. These dams could regulate the river flow, control floods, sedimentation and siltation, produce cheap hydropower, and reduce water losses by moving storage in areas with lower evaporation rates upstream (Guariso and Whittington 1987: 113; Allan 1994: 21; Tvedt 2010: 239; Blackmore and Whittington, 2008: ix). However, given its dependence on the Nile to meet most of its water needs and in light of the history of uneasy hydropolitical relations between Egypt and Ethiopia, Egypt has adamantly rejected the construction of large dams upstream that could affect its historical share of the Nile waters (Arsano 2007: 197–203). It was thus not surprising that the launching of the GERD increased the tensions between Ethiopia, that insisted that the project would cause no harm to downstream countries, and Egypt, that considered the project to be a threat to its water and national security (Tawfik 2015: 31–32).

The signing of the Declaration of Principles on the GERD by Egypt, Ethiopia and Sudan in March 2015 promised to ease these tensions. According to the Declaration, the three Eastern Nile countries would agree on the guidelines and rules of the first filling and annual operation of the dam. However, Ethiopia (as the owner of the project) retained the right to adjust these rules and 'inform' downstream countries of any 'unforeseen or urgent circumstances' that had led to such adjustment (Agreement 2015). The declaration foresees a permanent coordination mechanism to sustain coordination on the annual operation of the GERD, together with downstream reservoirs.

Against this background, the chapter examines the extent to which the GERD and the DoP have transformed, or can transform the hydropolitical interactions and debate in the Eastern Nile from sharing water resources to sharing benefits from projects constructed on the transboundary river. The second section briefly discusses the concept of benefit-sharing as a framework for analysing the benefits of cooperation and costs of non-cooperation in transboundary rivers, taking biophysical, economic and political dimensions into account. The third section applies this concept to the GERD, identifying its potential benefits and costs to Eastern Nile countries, and potential cooperation beyond the project. The fourth section examines the challenges that may face current and future negotiations over the GERD. Section five scrutinises the factors that are likely to affect these negotiations. Section six concludes by suggesting some measures to reach a benefit-sharing arrangement.

Cooperation in transboundary rivers: from water-sharing to benefit-sharing?

To assess the GERD's potential benefits and costs, this chapter applies the concept of benefit-sharing (Sadoff and Grey 2002, 2005; Dombrowsky 2009; Hensengerth, Dombrowsky and Scheumann 2012; Scheumann, Dombrowsky and Hensengerth 2014; Dombrowsky et al. 2014). The concept, as put forward by Sadoff and Grey, suggested broadening the scope of cooperation by transitioning from sharing water quantities to sharing benefits from the use of the river. The expectation is that this may help transcending differences over water allocations, which dominate in many transboundary rivers, especially the Nile River (Sadoff and Grey, 2008: 21).

Other scholars of benefit-sharing, however, suggested that sharing the benefits of cooperation may not always be an 'alternative strategy' to sharing water. As Dombrowsky (2009, 2010) argues, in cases where negative external effects are expected from a water use or project constructed in a riparian state, water allocation remains at stake. In line with the Coase Theorem, any negotiations on internalising negative externalities require clearly defined property rights as starting point of negotiation (Coase 1960). Hence, agreement on water rights might be a prerequisite for generating gains from cooperation or reducing the costs of non-cooperation.

According to Sadoff and Grey (2002), four types of benefits could be achieved through cooperation on shared rivers. The first type is *benefits to the river (environmental benefits)*, which may include enhancing water quality and flow, improving watershed management, reducing sediments transport and protecting biodiversity. The second type is *benefits from the river*, which refers to economic benefits from hydraulic projects constructed on the shared river. The social and environmental impact of such projects cannot, however, be ignored. The third type is *the reduction of costs because of the river (political benefits)*, which is associated with containing the tensions resulting from conflicts over water resources. The last type is *benefits beyond the river*. The argument is that: 'increasing the benefits from the river and decreasing the costs arising because of the river enable broader economic growth and regional integration that can generate benefits even in apparently unrelated sectors' (Sadoff and Grey, 2002: 11). These may include increasing trade relations, investments and hydropower interconnections.

Benefits would be maximised, if riparian states managed the basin 'holistically and efficiently, conserving essential ecosystems, and locating energy, industry, and agricultural development where it is most productive and least environmentally and socially disruptive' (Sadoff and Grey 2008: 22). Positive progress in one category of benefits may encourage advances in other types, but setbacks in one area can also hinder cooperation in others (Sadoff and Grey 2005: 3).

To facilitate cooperation, Sadoff and Grey (2005, 2008) recommend that the parties expand the perception of benefits to include more than one of the four types identified above and explore the distribution of benefits and costs in a way that is seen as fair to all riparian states involved through the use of the compensation, joint funding and ownership, and extracting benefits from unrelated projects. Riparian states can also examine the various modalities of cooperation, ranging from sharing of information to adaptation of national plans for mutual benefits and implementation of joint projects, to maximise benefits for all parties.

In this context, the perceptions of the riparian states of the extent to which the distribution of benefits is equitable are no less important than the generation of these benefits. As Sadoff and Grey noted, 'even significant gains of cooperation in a river system may not be sufficient motivation for cooperation if the distribution of those gains is, or is perceived as, inequitable' (2005: 396). In this case, cooperation will be less attractive, even if a riparian state perceives it as more beneficial compared to non-cooperation. The redistribution of different types of benefits and compensations are necessary means to ensure a balanced allocation of the returns of cooperation.

Scholars applying benefit-sharing have not only outlined the range of benefits that accrue from cooperation, but have also started to identify the mechanisms by which riparian states share benefits (and costs), as well as the incentives and factors that may influence a state's decision to engage in benefit-sharing negotiations and arrangements. Based on a number of case studies,

Hensengerth, Dombrowsky and Scheumann (2012: 5–25 and see also Scheumann *et al.* 2014) suggested that incentives for negotiating and concluding a benefit-sharing agreement on dam projects on transboundary rivers may include: (i) overcoming financial limitations to unilateral action (the Diama and Manantali projects on Senegal River basin); (ii) increasing collective benefits from a modified dam design (the Duncan, Keenleyside and Mica Dams on the Columbia River, British Columbia) or from locating a dam upstream (the Lesotho Highland Water Project on the Orange-Senqu River); (iii) receiving compensation for negative externalities and reducing the costs of conflicts (the High Aswan Dam on the Nile River); or (iv) producing mutual benefits from a joint dam on a border river (the Kariba Dam on the Zambezi River).

Mechanisms for benefit-sharing included sharing the financial costs between riparian states in proportion to benefits in return for joint ownership of the project (Senegal and Zambezi), compensation for modified dam design while sharing the benefits of cooperation (Columbia), or funding a project in another riparian state in return for sharing its benefits (Orange-Senqu). The authors assume that riparian states are willing to engage in benefit-sharing, if they believe they will be better off with cooperation than without.

The existence of (perceived) mutual benefits may, however, not be sufficient to reach an agreement. The latter is rather affected by many factors that are not confined to riparian states' national water policies and priorities. These factors might also include the foreign policies of the riparian states, the history of cooperation (or competition) between them, the political and administrative system within each riparian state, the role of third parties and regional organisations, and power relations between riparian states (Hensengerth *et al.* 2012: 26–30).

Applying this framework to the GERD, the next sections illustrate how the negotiations over the GERD will necessarily raise questions on the sharing of water, not only the sharing of the project's benefits. The sections examine the potential benefits and costs of the project to Ethiopia and downstream countries, the perceptions of the three riparians of these benefits and costs, the factors that affect cooperation to achieve these benefits and reduce costs, and the policy options available for the three Eastern Nile countries to increase the prospects of cooperation.

GERD from a benefit-sharing perspective

Ethiopia's unilateral launch of the GERD, the largest hydropower project on the Blue Nile with a storage capacity of 74 billion cubic metres (BCM) and a power generating capacity of 6,000 megawatt (MW), and its insistence to finance the project from domestic resources is a game changer in the Nile Basin. Although dam plans date back to 1964 when the United States Bureau of Reclamation (USBR) concluded comprehensive studies on the utilisation of the Blue Nile, Ethiopia lacked the political, diplomatic, technical and, more importantly, economic capacity to implement these plans (Allan 1999: 4; Swain 2002: 298;

Waterbury 2002: 68; Arsano 2007: 90). Since 1999, Ethiopia has tried through its participation in the Nile Basin Initiative (NBI), the intergovernmental mechanism of cooperation in the basin, to mobilise financial resources for the updating and joint implementation of these plans.

In the framework of the NBI, the Eastern Nile Technical Regional Office (ENTRO) hired a consultant to conduct the Eastern Nile power trade study, which included pre-feasibility studies for the Mandaya and Border Dams on the Blue Nile. Mandaya Dam was planned to be a hydropower project with an installed capacity of 2400–2800 MW. The Border Dam, with a storage capacity of 14.5 BCM and an installed power capacity of 800 MW, was supposed to be launched in the same location where the GERD is now being constructed. The pre-feasibility studies of the two projects were finalised in 2008. Meanwhile, the Ethiopian government signed contracts with a consortium of Norwegian companies to conduct the pre-feasibility study for the Karadobi Multi-Purpose Dam and the feasibility study of the Beko-Abo Dam, two additional projects on the Nile tributaries (Ministry of Foreign Affairs of the Arab Republic of Egypt 2014b; Ministry of Water Resources of the Federal Democratic Republic of Ethiopia 2007).

However, the decision of most upstream countries in 2010 to sign the Cooperative Framework Agreement for the Nile Basin (CFA), negotiated since the end of the 1990s, before reaching consensus on all its articles, and the subsequent decision of Egypt to suspend its membership in the NBI, have affected the implementation of joint projects in the framework of the initiative (Tawfik 2015: 18–20). It was, thus, unsurprising that Ethiopia proceeded with the construction of the GERD outside the framework of the initiative (See Cascão and Nicol, Chapter 5). The following section will first discuss potential benefits from and to the river in relation to GERD, second, whether GERD helps reducing costs because of the river, and third, whether it motivates benefits beyond the river.

GERD: *uneven benefits from and to the river, differentiated incentives*

The GERD is central to Ethiopia's development vision of becoming a middle-income country by 2020–2023 (Ministry of Finance and Economic Development of the Federal Democratic Republic of Ethiopia 2010) and Africa's energy hub. The GERD is aimed primarily at raising hydropower production for both the domestic and regional markets (Zenawi 2011). Ethiopia's electricity exports are expected to increase tremendously once the GERD is commissioned, securing more hard currency for the Ethiopian government. However, extending transmission lines and expanding the export of hydropower need to accrue as quickly as possible, otherwise the realization of the expected revenues could be significantly delayed (Massachusetts Institute of Technology 2014: 8; Whittington *et al.* 2014: 600). Still, at least in the medium to long term, Ethiopia expects significant economic benefits from the project through the export of energy, as such generating benefits even 'beyond the river'. To make sure that benefits from the dam will be delivered once it is commissioned, the Ethiopian government signed a deal with a Chinese company in April 2013 to connect the GERD to the national grid before the dam is completed (Addis Fortune 2013). In October 2016, Sudan announced that it would build a new transmission line from the GERD to link Ethiopia and Sudan electricity networks (Sudan Tribune 2016).

Environmentally, Ethiopia's gradual dependence on hydropower would reduce the environmental and health impacts of biomass fuel used by most of its population (Hammond 2013: 2; King 2013: 2). The project's Environmental and Social Impact Assessment (ESIA) so far was only presented to the International Panel of Experts (IPoE), a panel comprised experts from the three Eastern Nile countries, in addition to international experts, that was set up after the launching of the project to examine its documents, but not made public. In its final report, the IPoE noted that the ESIA outlined the positive and negative physical, biological and socio-economic impacts of the project on the Blue Nile within Ethiopia. It further identified the means and financial costs of mitigation measures (IPoE 2013: 18).

But the GERD has also sparked controversy about its financial cost compared to its returns. According to the Ethiopian Minister of Communication and Technology and the deputy Prime Minister, Debretsion Gebremichael, out of an expected total cost of US$4.8 billion the dam has consumed 46 billion birr (around US$2.2 billion) until October 2015, 16 per cent of which were collected from selling bonds to the public at home and in the diaspora (Strategic Thinking on East Africa 2015). The International Monetary Fund (IMF) advised the Ethiopian government to slow the construction of the GERD to avoid the dam absorbing most of the domestic finances. According to the Fund (IMF 2014: 6–7), the Ethiopian government has borrowed heavily from national banks to invest in the GERD and other large infrastructure projects, which could undermine the country's macroeconomic stability.

The justification of the dam's financial cost in relation to its efficiency in power production and its overall economic potential has also been debated. The report of the IPoE noted that although the project appears to be economically attractive, no economic justification was given by the Ethiopian government for the installed power capacity of 6,000 MW (IPoE 2013: 38). Other sources suggested that the dam could not possibly produce the targeted power production throughout the year because it was designed for a near peak flow (Beyene 2013). The IPoE report (2013: 35) also pointed out that although the estimations of sediment yield and trap efficiency in the GERD's reservoir were realistic, sediment accumulation over time has not been considered, a factor that affects the lifespan and economic viability of the project (Chen and Swain 2014: 14–15; International Rivers 2012: 39; Veilleux 2013: 12). Equally important, climate change data has to be integrated in the planning of the project, a variable that the Ethiopian government ignored (IPoE 2013: 35).

As far as downstream countries are concerned, the GERD has different potential impacts that vary according to the sector and country. According to the IPoE, the Ethiopian government designed the dam based on technical criteria (power production and reservoir filling), but hardly took downstream impacts into consideration. It is against this background that the IPoE (2013: 40–42) recommended 'a more comprehensive assessment' that takes into consideration the project's benefits and costs to downstream Egypt and Sudan.

The IPoE report and other studies that applied different models to anticipate the impact of the project on downstream countries suggest that the overall net benefits to Sudan could be positive. The regulation of the flow would potentially increase the irrigated agricultural area in Sudan, a major benefit to Sudan, but a source of a serious concern to Egypt, as will be illustrated later in this section (see also Cascão and Nicol, Chapter 5). It is also expected that the GERD would reduce sedimentation and, thus, improve the operation of Sudanese dams and reduce the financial cost of rehabilitating these dams (IPoE 2013: 36; Kahsay et al. 2015: 6). According to the NBI's *State of the River Nile Report* (2012: 49), the storage capacity of the Roseires and Khashm el Girba reservoirs have fallen by 60 per cent, and 40 per cent respectively over the last thirty years. Additionally, (and 'beyond the river') Sudan would also benefit from the imported cheap electricity produced by the dam. These benefits may explain Sudan's political support for the project (Ministry of Foreign Affairs of the Republic of Sudan 2014b). In March 2012, even before the IPoE finalised its report on the dam, the Sudanese President Omar al-Bashir, announced while receiving the new Ethiopian Ambassador to Sudan that his government recognised the mutual benefits of the project to Sudan and Ethiopia and that he would extend his support to ensure its successful completion (Sudan Tribune 2012).

However, the GERD has also potential negative impacts on Sudan that the Ethiopian government, as noted in the IPoE report, neglected. These include the dam's impact on Roseires' biodiversity and fisheries, riverbed and bank erosion, groundwater levels, recession agriculture and the brick industry (IPoE 2013: 41; Mohamed, 2015). As far as water quality is concerned, the IPoE (2013: 39–42) noted that the reduction of dissolved oxygen because of the degradation of eroded vegetation and soil in the reservoir was not adequately addressed. The deterioration of water quality would affect aquatic biodiversity and fisheries at the reservoir and downstream (International Rivers 2012: 13–16).

The dam's potential negative impacts on Egypt would largely depend on the dam's filling and operation strategies, and thus on ongoing negotiations between the three countries (see Wheeler, Chapter 10). For its part, the Egyptian government claimed that it had conducted its own transboundary impact assessment and concluded that the dam 'would cause appreciable harm, including material environmental and socioeconomic harm to Egypt' (Ministry of Foreign Affairs of the Arab Republic of Egypt 2014b), without publishing this assessment or explaining what negative impacts were expected and how they could be mitigated. The IPoE (2013: 36) suggested that the water flow to Egypt would not be affected during the first filling in case it was done in a wet or average year, but power generation at the High Aswan Dam (HAD) would be cut by about 6 per cent. Should the filling occur during dry years, water supply and power generation in Egypt would significantly be reduced. However, assuming that the HAD would be operated at lower levels in the long run, scholars suggested that water losses from evaporation at the HAD would be reduced, meaning that there would be net water gains in the system. This, according to the authors, could possibly increase irrigation water supply to Egypt in the dam's long-term operation phase, while reducing power production from the HAD in the long term (Mulat and Moges 2014; Kahsay et al. 2015: 6; and see also Kahsay et al., Chapter 8).

Additionally, any assessment of the transboundary impact of the GERD on water flow to Egypt has to factor the potential increase in Sudan's water withdrawals for irrigated agriculture after the operation of the GERD (IPoE 2013: 42; Kahsay et al., 2015: 6). According to the 1959 bilateral agreement for the utilisation of the Nile water, 18.5 BCM of the Nile water at Aswan was allocated to Sudan and 55.5 BCM to Egypt. Until recently, Egypt has been using Sudan's unutilised share. With the expansion of hydraulic projects in Sudan during the last decade, Egypt started to raise concerns that these new projects would tempt Sudan to use more than its share (Cascão 2009: 257–258). The Egyptian government expressed these concerns in discussions with donor countries and institutions (Abul Gheit 2014: 267). The GERD feeds these concerns and affects Egypt's position on the project.

More benefits could even be realized, if a basin-wise planning approach was taken. In line with the concept of benefit-sharing which stresses the importance of holistic planning of transboundary rivers, the NBI's *State of the River Report* emphasises the need to develop basin-wide development plans (or at least sub-basin plans) that encompass all water uses by riparian states and the cumulative effect of several projects that are planned or under construction (NBI 2012: 175). The NBI and its Eastern Nile Technical Regional Office (ENTRO) seek to face common challenges, including climate change, restoration of degraded water catchments and the establishment of a regional hydrometric and environmental monitoring system (NBI 2012: 26, 212–215). But these challenges seem to have been relegated to a secondary priority for Eastern Nile countries after the tensions over the CFA and the subsequent Egyptian decision to suspend its membership in the NBI in 2010, and since then, the construction of the GERD has dominated the Eastern Nile agenda.

To sum up, the three Eastern Nile countries have incentives to coordinate on the GERD since its transboundary impacts would depend on the filling and operation strategies. However, since the dam holds uneven potential benefits and costs to the three Eastern Nile countries, negotiation over a potential redistribution of benefits, the compensation for losses and the potential generation of other benefits from other projects would be necessary to reach benefit-sharing arrangements that would be seen by the three countries as fair and equitable. The different estimations of the downstream impacts of the GERD also suggest that joint planning, monitoring and operation of the dam would be desirable (Kahsay et al. 2015: 14; Mulat and Moges 2014: 591). This is especially the case since Ethiopia will be interested in filling the reservoir more quickly to maximise power generation, while downstream countries, especially Egypt, would prefer slow filling to reduce potential reduction of water flow (MIT 2014: 4–5). These issues will be discussed in detail below. Before doing so, we discuss how the GERD relates to the costs of non-cooperation and benefits beyond the river.

GERD: *reducing costs of non-cooperation or increasing political tensions?*

Rather than reducing the costs of non-cooperation, at least initially, the unilateral launch of the GERD increased political tensions, especially between Egypt and Ethiopia. Ethiopia presented the GERD as a sign of its determination to redress historical injustices in the utilisation of the Nile water resources and exercise its right to use these resources, even if it involved high financial costs (Zenawi 2011). It is also seen as a 'unifying force across ethnically diverse and divided Ethiopia', and as a key to changing Ethiopia's image from a country of famine and dependence on aid to a strong, self-sufficient economy (Veilleux 2013: 8).

For its part, Egypt sees the project as a 'significant threat to its national and water security' (Ministry of Foreign Affairs of the Arab Republic of Egypt 2014b; Fahmy 2014). After the completion of the dam, and given its large storage capacity, the flow of water to Egypt will depend on Ethiopia's cooperation in providing sufficient annual releases. In this sense, the GERD exposes Egypt's vulnerability associated with its high dependence on a water resource that originates outside its borders. Additionally, a number of Egyptian experts and scholars considered the Ethiopian decision to increase the storage capacity of the dam (compared to earlier plans) as an indicator that the project is more about controlling the flow of the Nile than producing electricity (Nour Eddin 2014; Raslan 2013; Sharaqi 2011).

From the launching of the project until May 2013, Egyptian successive governments responded to the project by diplomatic means, calling upon Addis Ababa to respect the principles of international law, especially the obligation not to cause significant harm, and the principles of prior notification and consultation on projects that would affect other riparian states and of sharing information. The Ethiopian decision to divert the Blue Nile on 28 May 2013 to commence the construction of the dam's body, a few days before the IPoE had submitted its report invited Egypt to take a tougher stance. The former Egyptian President Mohamed Morsi threatened to use force if Egypt's share of the Nile water was affected, insisting that Egypt would accept no infringements on its water security (Ahram Online 2013).

The ouster of Morsi in July 2013 only partially reduced these tensions. In January 2014, Egypt withdrew from the ministerial rounds of negotiations over the implementation of the IPoE report, declaring its frustration from the failure to reach an agreement that secures the interests of the three countries, and from the Ethiopian rejection of suspending the construction until reaching this agreement as a confidence-building measure (Ministry of Foreign Affairs of the Arab

Republic of Egypt 2014b). This was followed by an official statement that Egypt is preparing a comprehensive legal file about the Renaissance Dam, although its primary goal remained to engage in negotiations to solve the dispute (Ministry of Foreign Affairs of the Arab Republic of Egypt 2014c). Beyond this statement, the legal track was not, however, pursued by the Egyptian government.

On another track, to balance the Ethiopian–Sudanese rapprochement and convergence of positions on the GERD, Egypt sought to develop its relations with other Nile Basin countries, especially South Sudan in areas that are not confined to water cooperation. This included signing a military agreement with Juba in March 2014 to train the South Sudanese army in Egyptian military colleges (Al-Monitor 2014). The agreement raised such concerns in Ethiopia that South Sudan's President Salva Kiir had to assure the Ethiopian prime minister that the agreement was not directed at Addis Ababa and its Renaissance Dam (The Reporter 2014; World Bulletin 2014). The agreement came after the Ethiopian government's decision to set up a joint military force with Sudan 'to secure borders' and 'boost the opportunities of joint development' (Ministry of Foreign Affairs of the Republic of Sudan 2014a).

It was only in August 2014, after the election of a new president in Egypt, that technical negotiations were resumed to deliberate the implementation of the IPoE's recommendations. This paved the way for the signing of a Declaration of Principles (DoP) between the three Eastern Nile countries on the GERD in March 2015. The DoP has reduced, but not ended, the political tensions over the GERD. The accusations by the Ethiopian government that Egypt is supporting the Ethiopian opposition that orchestrated the demonstrations in the Oromia region, in late 2016, indicate the low level of trust between the current regimes in Cairo and Addis Ababa in spite of the recent rapprochement (Daily News Egypt 2016). These accusations echoed long-standing claims by the Ethiopian government that Egypt promotes instability in Ethiopia and the Horn of Africa to prevent it from utilising the Nile waters (Ministry of Information of the Federal Democratic Republic of Ethiopia 2002: 120).

For the same claims, Ethiopia continues to watch closely and uneasily Egypt's move to foster relations with other Nile Basin countries. It showed reservation on Egypt's proposal to contribute to the United Nations peacekeeping mission in South Sudan. In response to the Egyptian plans, the former Ethiopian Minister of Communication, Getachew Reda, indicated that Ethiopia would not be worried if the Egyptian troops in South Sudan are 'exclusively for peacekeeping purposes', adding that 'his government will do something if there is any effort whatsoever by the Egyptians to sabotage Ethiopia's success' (The Reporter 2016). A recent Egyptian presidential visit to Uganda, followed by a visit by South Sudanese President Salva Kiir to Egypt raised concerns in Addis Ababa that Kiir had to publicly dismiss claims that three countries conspire against Ethiopia (The East African 2017).

Hence, at this point, it remains unclear how the GERD will affect political relations in the long term, what is clear, however, is that the GERD is a game changer of the hydropolitical relations in the Nile Basin.

Beyond the river: furthering or hindering regional integration?

The GERD promises to open avenues of cooperation beyond water resources, especially in energy trade between Nile Basin countries. The expected increase in Ethiopian electricity export to Kenya, South Sudan and Sudan would significantly increase the volume of power trade in the Nile Basin, which is currently insignificant compared to other sub-regions in Africa (NBI 2012: 164). In this sense, the GERD, and the DoP, are steps towards the achievement of the Ethiopian vision for 'regional integration through the generation of sustainable and clean energy supply' (Agreement 2015: Article II). This vision promotes Ethiopia's regional position as an energy hub, while encouraging other Eastern Nile countries to specialise in other areas of comparative advantage – industrial development for Egypt and agriculture for Sudan (Arsano 2007: 226; Ministry of Information of the Federal Democratic Republic of Ethiopia 2002: 123–124).

The Ethiopian–Sudanese rapprochement and convergence of positions on the GERD may help in achieving this vision and represent a turning point in hydropolitical relations in the Nile Basin. With the signing of the DoP, Egypt tried to catch up with Ethiopia and Sudan's integration endeavours. One day after the signing of the Declaration in Khartoum in 23 March 2015, the Egyptian President Abdel Fattah El-Sisi paid a state visit to Ethiopia. During the visit, he agreed with the Ethiopian Prime Minister, Hailemariam Desalegn, to elevate the bilateral ministerial commission to the presidential level to enhance bilateral cooperation (State Information Service of the Arab Republic of Egypt 2015). This was followed by a number of steps to boost Egypt's relationships with Sudan in the fields of agriculture, industrial investment promotion and road connection (Ahram Online 2016). During the Africa Business Forum in Sharm el-Sheikh in February 2016, the leaders of Egypt, Ethiopia and Sudan agreed to set up an investment fund to mobilise finances for joint projects. However, one year after the summit, the fund is yet to materialise (Ministry of Foreign Affairs of the Federal Democratic Republic of Ethiopia 2016). Moreover, Egypt continues to link its cooperation in certain fields with Nile Basin countries in general, and Eastern Nile countries in particular, to agreement on the utilisation of Nile water resources. The Egyptian refusal to sign the master plan of the East African Power Pool (EAPP) adopted by Burundi, The Democratic Republic of Congo (DRC), Ethiopia, Kenya, Rwanda, Sudan, Tanzania, Libya and Uganda, in the EAPP's 2016 meeting in Addis Ababa, citing concerns over Nile projects is a case in point (The East African 2016).

To sum up, although the hydropolitical interactions around the GERD have moved towards more coordination, the project has not transformed the debate around the utilisation of water resources in the Eastern Nile from sharing water to sharing benefits. Rather than constructing the dam as part of a holistic plan for the management of the transboundary river, as the benefit-sharing scholars recommend, political tensions over the CFA and the freezing of joint projects have hindered the implementation of the GERD within a more comprehensive plan of water resources development in the basin. Furthermore, the implementation of the project without comprehensive assessment of its transboundary impacts and the increase in the dam's size and storage capacity compared to earlier plans have increased the economic and political costs of the project. As a result, negotiations over the last five years have focused more on conducting the required studies on the impact of the project, than on maximising its benefits and actually reducing its potential negative impacts, in particular those during the filling period. The maximisation of benefits would have presupposed a halt in construction and an open-ended discussion on detailed project design. Rather than seeing the dam as a tool for a better management of the river and a driving force for regional integration, Egypt remains concerned about the impact of the project on what it considers its historical share of the Nile water. This concern has affected Egypt's position on Eastern Nile political and economic cooperation. While the three Eastern Nile countries have incentives to coordinate on the project, its uneven potential benefits to them mean that negotiations over the redistribution of benefits and compensation for losses would be necessary to induce them to implement the agreements reached on the GERD and foster cooperation beyond the project. The next section explores challenges in coming up with a benefit-sharing deal that is seen as fair by the three countries is possible.

GERD: challenges in moving towards a benefit-sharing deal

In September 2016, more than one year after the signing of the DoP, Egypt, Ethiopia and Sudan signed the contracts for the technical studies on the downstream impacts of the GERD, which will take eleven months to be finalised. Tough talks are expected when the guidelines of filling and operation of the dam will be negotiated. Three basic issues are often contentious in negotiations over the generation and sharing of benefits in transboundary rivers, and are expected to be even more disputable in light of the historical hydropolitical relations and ongoing geopolitical transformations in the Nile Basin. These issues include: (1) defining the starting point for negotiations; (2) negotiating significant harm and compensation; and (3) agreeing on the powers of the coordination mechanism for annual dam operation. These issues indicate that sharing the benefits and dealing with the potential negative impacts of the project will necessarily prompt the three countries to negotiate water shares.

The first issue relates to the basis or starting point for negotiation, which is related to whether there are clearly defined property rights (Dombrowsky 2009: 128–131). The DoP provides general rules to guide the negotiations, namely, the principles of the equitable and reasonable utilisation of shared water resources and of causing no significant harm (Agreement 2015: Arts III and IV). But given the absence of an agreement for water allocation that is accepted by all riparian states, the prioritisation of the factors defining the equitable and reasonable utilisation and the definition of significant harm would be disputable. Egypt is expected to take its historical share of the Nile waters as a reference point in defining significant harm. However, the DoP, in line with the United Nations Convention on the Law of Non-Navigational Uses of International

Watercourses, lists existing, together with potential future uses, as relevant factors in the definition of the equitable and reasonable utilisation (UN Convention 1997: Article 5). The Egyptian Ministry of Foreign Affairs (2015b), however, announced in a statement issued two days before signing the declaration that the DoP would not affect historical agreements and the water share allocated in the 1959 agreement. This statement is consistent with the previously adopted Egyptian constitution that commits the state to 'preserve Egypt's historical rights' in the River Nile (Constitution, 2014: Article 44). Sudan remains officially committed to the bilateral 1959 agreement with Egypt, which allocates 18.5 BCM to it annually. Still, as mentioned above, the potentially increasing water withdrawals by Sudan after the completion of the GERD, and whether these could raise Sudan's *actual* withdrawals beyond its allocated share, need to be assessed.

Accepting the current uses as a starting point would mean that Ethiopia recognises the historical shares of Egypt and Sudan, a compromise by Ethiopia that would lead to filling and operating the dam to secure these shares. It would also mean that Sudan would continue to be committed to the 1959 agreement. However, it can be argued that for Ethiopia accepting historical shares as a starting point is not fully in the spirit of the DoP, which refers to existing and potential uses to be considered for an equitable and reasonable utilisation. Neglecting these historical shares would allow Ethiopia to fill and operate the dam to reap maximum benefits, and/or allow Sudan to withdraw water beyond its share, but entail a significant harm to Egypt. This, however, would also not be in the spirit of the DoP. A middle ground could be considering existing and potential uses and reaching a compromise for the starting point of the negotiations.

The second issue relates to the negotiation of significant harm and compensation. The DoP committed Ethiopia to mitigate significant harm in case it occurs and discuss compensation to downstream countries (Agreement 2015: Article VI). The definition of significant harm arguably again depends on agreeing what constitutes an equitable and reasonable utilisation, as the two principles are closely related. If the current allocations under the 1959 Agreement were accepted as a starting point, it would mean that Ethiopia accepts to pay, and that Egypt accepts to receive, compensation for any significant harm that may affect the latter's historical share. If equitable and reasonable utilisation was defined on the basis of existing and potential uses, this would alter the definition of significant harm. However, Ethiopia could theoretically also ask for compensation for adapting the filling and annual operation to the demands of the downstream countries. The DoP is largely, and may intentionally be, silent on this issue, besides the fact that it gives Ethiopia the right to adjust the agreed rules of operation 'from time to time' (Agreement 2015: Article V).

A third issue relates to the powers of the coordination mechanism set by the DoP to sustain coordination on the annual operation of the GERD, together with existing downstream reservoirs. This proposed trilateral mechanism seems to be more of a coordinating instrument to facilitate agreement on the rules of filling and operation than a supranational authority that ensures the implementation of this agreement. This is compounded by the fact that the Ethiopian government as the owner of the project has the right to change the guidelines of operation in 'unforeseen or urgent circumstances' (Agreement 2015: Article V) without prior consultation with downstream countries, a clause that may open the door for disagreements in the future and augment Egypt's vulnerability to the Ethiopian government's management of the dam. In this context, the responsibilities of the coordinating mechanism in monitoring the operation of the GERD and of Sudanese dams need to be clearly demarcated. A greater commitment by Ethiopia, as the owner of the project, on setting up a permanent coordination mechanism could again play a role in coming to compromise on equitable and reasonable utilisation.

These three issues indicate the complexity of the ongoing negotiations on the GERD. They confirm that next to benefit-sharing also water allocations are at stake and that sharing benefits is not always an alternative strategy to sharing water (Dombrowsky 2009, 2010). They further suggest that generating additional benefits from the GERD and from related and/or unrelated projects may be necessary to induce different parties to offer concessions in negotiations. Ethiopia will need to market the hydropower produced by the project. It is likely to face regional competition with the entry of other Nile Basin countries, including the DRC and Uganda, to the power trade market (Matthews et al. 2013). Here, Egypt could help and could link this assistance in marketing hydropower produced by the project to Ethiopia responding to downstream demands in filling and operating the dam. Although the DoP gives Egypt and Sudan priority to purchase power generated from the GERD (Agreement 2015: Article VI), Egypt, contrary to Sudan, has not voiced its interest in hydropower produced by the project. A panel of experts formed by the Egyptian Ministry of Water Resources and Irrigation to respond to a report issued by the Massachusetts Institute of Technology rather confirmed that Egypt would not sign a hydropower agreement with Ethiopia, if guarantees of causing no significant harm to Egypt were not secured (MIT 2015, 6). More recently, the Ethiopian government announced that it has finalised a feasibility study on electricity export to Egypt and that consultations are underway between the two countries to supply power to Egypt (Ethiopian News Agency 2017). The Egyptian government, however, responded by indicating that Egypt is keen on cooperating with Nile upstream countries in the field of energy and power trade, but importing electricity from Ethiopia is unlikely in the meantime. It further emphasised that Egypt plans to develop its role as an energy hub and exporter of electricity to other countries in the Middle East (Al-Masry Al-Youm 2017). Purchasing and marketing hydropower generated from the GERD could thus induce Ethiopia to offer compromises in negotiations. Increasing virtual water trade imports in livestock and crops from Ethiopia and Sudan to Egypt, and Egyptian investments in the agricultural sector in the two upstream countries could generate benefits 'beyond the river' that foster regional integration and reduce tensions over sharing water resources.

Sharing the costs of the annual operation and maintenance of the GERD could provide another incentive for Ethiopia to cooperate. Beyond the GERD, coordination to mobilise international funding for future projects to increase the river's discharge and implement joint developmental projects in the Eastern Nile would generate more benefits to the three countries. A number of factors will affect reaching compromises on these and other issues. These are discussed briefly below.

Factors affecting the conclusion of a benefit-sharing deal

Various factors may affect the conclusion of a benefit-sharing deal over the GERD. These factors are not confined to the water policies of the three Eastern Nile countries, but also include their foreign policies, domestic power structures, the role of regional organisations in building trust and power relations between the three riparian states.

Foreign policies of the three governments: can they overcome historical mistrust?

It has been argued that cooperative foreign policies and the willingness of the riparian states to build trust in areas that are not confined to the water sector would facilitate negotiations over shared water resources (Hensengerth *et al.* 2012: 27). Although the three Eastern Nile countries have moved from confrontation to coordination over GERD and beyond, mistrust continues to prevail. In spite of improvements in political and economic relations between Ethiopia and Egypt, tensions over the GERD continue to dominate the agenda of bilateral cooperation. Beyond what has been indicated above already, relationships between Egypt and Sudan continue to be shaky. The recent Sudanese complaint to the United Nations Security Council against Egypt for holding elections in the disputed Halayeb Triangle, and its contestations of Egypt's disrespect of the signed bilateral agreements of free movement of goods and persons in general, and of the rights of Sudanese in Egypt in particular, are clear indicators (The New Arab 2015).

Simultaneously, Addis Ababa and Khartoum are taking serious steps towards cooperation. The two countries have recently agreed to increase the use of Port Sudan to serve the northern part of Ethiopia, raise funds to connect the port with Ethiopia by railway, complete the high electric transmission lines for energy trade and establish a free trade zone along their border (Ethiopian New Agency 2015). This may encourage Egypt to follow suit to foster Eastern Nile integration, or augment its sense of isolation vis-à-vis the other two countries. On one hand, the Egyptian government seems to recognise the high political cost of non-cooperation and disengagement. Stressing the importance of negotiations on the GERD and taking steps to increase cooperation with Ethiopia and Sudan beyond the project reflect this recognition. At the same time, its position towards Eastern Nile integration seems cautious and conditional on progress in GERD negotiations. The Egyptian position towards the EAPP energy connection illustrated in earlier is a case in point.

Domestic power structures and decision-making processes

A riparian state's position in negotiations over shared water resources or dam projects constructed on a transboundary river is affected by the coherence of its domestic institutions and the influence of public opinion. The multiplicity of institutions involved in the Nile issue in each of the Eastern Nile countries, the shortage of coordination between these institutions and the lack of transparent decision-making may complicate the process of negotiation. This is particularly evident in Egypt whose position on the GERD has been far less consistent than Ethiopia's and Sudan's. Therefore, the following will mainly focus on Egypt to illustrate the point. The Egyptian policy on the GERD has been managed by various institutions, including the Ministry of Water and Irrigation, the Ministry of Foreign Affairs, the Presidency and the national security agencies, with no clear division of labour between them.

An example of this problem is the confusion that dominated before the signing of the DoP. A few days before signing the Declaration, state-owned media sources in Egypt reported that the High Committee on the Nile Water, a committee led by the prime minister and comprising the ministers of water resources and irrigation, agriculture, foreign affairs, defence, international cooperation and the general intelligence chief, was considering renegotiating some of the formulations of the document before signing (Ahram 2015; Ministry of Foreign Affairs of the Arab Republic of Egypt 2015a). The signing of the Declaration by the president himself, without any revision, took many by surprise. The author's informal discussions with some Egyptian diplomats and technocrats reveal that some voices within the ministries of foreign affairs and water resources called for a refraining from signing the declaration. For them, signing the DOP meant that Egypt would formally accept the GERD without legally committing Ethiopia to respect Egypt's historical share. Instead, Egypt should, according to their view, have continued to contest the project, at least diplomatically, until it restored its power position.

This position indicates that divisions exist inside Egyptian policy circles on dealing with the GERD and Ethiopia's future dam projects. Given this division, it is unclear which institutions (or individuals) are most influential in decision-making, how final decisions are made and what the rationale of these decisions is. This makes it difficult to predict Egypt's negotiating policy and final decisions.

At the same time, the political sensitivity of the Nile issue in the three countries and the nationalistic rhetoric used by their governments, especially in Egypt and Ethiopia, may make offering concessions in future negotiations more difficult. So far, the Egyptian government has continued to assure Egyptians that the DoP, and any future arrangements, will not threaten its 'acquired rights' in the Nile water (Ministry of Foreign Affairs of the Arab Republic of Egypt 2015b). The Egyptian leadership stresses that the concerns of the Egyptians regarding the GERD are legitimate, emphasising that 'the Nile waters is a matter of life and death that no one can tamper with' (Daily News 2017).

Critical voices invite the Egyptian government to withdraw from negotiations and use other means to defend its historical rights. A group of Nile experts led by former Minister of Water Resources and Irrigation from March 2009 to January 2011, Mohamed Nasr Eddin Allam, suggested that Egypt should declare the failure of the ongoing technical negotiations, propose a suspension of the project until the conclusion of the required studies with a commitment to compensate Ethiopia for this delay, if the project proved to cause no significant harm to downstream countries, and resort to the UN Security Council, if Ethiopia refused this proposal (Nour Eddin 2015). Similar voices on the Ethiopian side call upon Addis Ababa not to 'surrender its sovereignty' and allow Egyptian intervention in Ethiopia's developmental projects (Negash *et al.* 2015; Gebremedhin 2015). There is no indication that the Egyptian and Ethiopian governments are preparing public opinion to any concessions in future negotiations.

National water policies and priorities

Reaching compromises on the GERD partly hinges on the willingness of the three Eastern Nile countries to adapt their water uses and policies to respond to the post-GERD era. In Egypt, a strategy for water resources until 2050 that focuses on developing alternative water resources, improving water quality, increasing water efficiency and strengthening the legal framework governing water utilisation was recently designed. It focuses on setting new projects for storing rainwater in the Northern coast, expanding desalination of seawater and groundwater projects and reducing the waste of irrigation water. Based on this strategy, a National Water Resources Plan that covers the period from 2017 to 2037 is currently being prepared with the support of the European Union (European Union 2016).

However, new mega-developmental projects, including land reclamation schemes, raise doubts about Egypt's management of limited alternative water resources, and thus its position towards any developments upstream that could affect its historical share of the Nile water. Particularly noteworthy is the government's plan to reclaim 1.5 million feddans in different locations starting from the Western Desert, with 11.5 per cent of the required water coming from the Nile, and the rest from groundwater (Ministry of Water Resources and Irrigation of the Arab Republic of Egypt 2016).

Sudan, the Nile riparian state with the second largest irrigated area in Africa after Egypt, is increasingly dependent on agriculture after the 2011 secession of the South, where the vast majority of oil reserves are located. Its twenty-five-year strategy (2002–2027) projects the water needs for irrigation to be about 42.5 BCM by 2027 and the total demand to be 59.2 BCM (Abdalla and Mohamed 2007: 4). Although the majority of agricultural lands in Sudan are rain fed, new irrigation schemes on the White and Blue Nile banks, funded either by the Sudanese government or by foreign investors, are withdrawing increasing volumes of water (Cascão 2009: 259). As noted earlier, more withdrawals from the Blue Nile Basin would be made possible after the regulation of flow with the commission of the GERD.

For Ethiopia, the GERD is likely not to be the last dam on the Nile. Furthermore, if the Ethiopian government was to follow the recommendation of its Water Strategy adopted in 2001 by focusing on multi-purpose dams (Ministry of Water Resources of the Federal Democratic Republic of Ethiopia, 2001: 8), future dam projects would not only be for hydropower production, but also for irrigation, thus increasing consumption of the Nile water. This will increase withdrawals from different Nile sub-basins and add new tensions to hydropolitical relations with Egypt.

Regional institutions and organisations

Cooperation between the riparian states in the framework of regional institutions and river basin organisations often helps to promote mutual confidence and reduce transaction cost (Hensengerth et al. 2012: 29). In the first decade of the millennium, the NBI and its ENTRO have played a positive role in building trust and encouraging coordinated planning of dams and other infrastructure through its Joint Multipurpose Programme (JMP). In June 2010, Egypt and Sudan announced freezing their membership in the NBI to contest the signing of the CFA by upstream Nile riparians. Sudan returned back to the NBI after two years, chaired its Council of Ministers (ENCOM) in 2014–2015, and actively participated in mobilising funding for the initiative's joint programmes.

For its part, Egypt has not yet resumed its membership in the initiative. The failure of the meetings held in March 2017 to discuss the return of Egypt to the NBI, to reach an agreement on addressing Egypt's concerns regarding the CFA, has widened the rift not only between Cairo and the river basin initiative, but also between Cairo and other riparian states. The lack of agreement on principles to govern the construction of hydraulic projects and the definition of water security will further complicate the agreement on related issues in GERD negotiations.

Power relations between Eastern Nile countries

It has been argued that asymmetrical power relations are less conducive to cooperation in transboundary rivers and to reaching benefit-sharing deals. Hydro-hegemony scholars cited Egypt's powerful position as a reason for the domination of a system of 'consolidated control' with little competition and a limited degree of cooperation (Zeitoun and Warner 2006: 437–438). Benefit-sharing theoreticians used the same factor to explain the lack of benefit-sharing projects in the Eastern Nile Basin as Egypt – having constructed the HAD and prior to the GERD – did not need to cooperate with upstream Ethiopia in constructing hydraulic projects, even if net benefits would increase through cooperation (Hensengerth et al. 2012: 29–30).

As indicated above, the GERD is changing hydropolitical relations on the Eastern Nile. The question is whether the GERD leads to more symmetric power or whether it reverses power relations in the Nile (Tawfik 2016). Egypt's declining economic power after five years of instability in the aftermath of the 25 January 2011 revolution, its military operations against terrorist groups in Sinai, and its diplomatic and military engagement in regional conflicts, in contrast with the relative stability and economic rise of Ethiopia and the increasing geopolitical influence of Sudan (See Cascão and Nicol, Chapter 5), are creating new power imbalances in the basin. For the first time, Ethiopia is able to combine the geographic power as an upstream country, the material power of sustained economic growth and the political power of its rising influence in the Horn of Africa. Based on the argument of the hydro-hegemony scholars cited above, it can be argued that in spite of the economic burdens imposed by the GERD, Ethiopia has fewer incentives to cooperate compared to Egypt. This will likely affect the prospects of reaching a deal between the three Eastern Nile countries and the content of this deal.

Conclusions

The GERD is presented by Ethiopia and by a number of scholars as a benefit-sharing project in the Nile Basin in general, and the Eastern Nile in particular. The resumption of talks in August 2014 and the signing of the DoP in March 2015 have reduced the tensions over the project and have paved the way for arrangements to share benefits, especially power produced by the dam, and compensate for potential negative impacts. The continuation of the negotiations in spite of difficulties indicates that the three governments believe that they will be better off with cooperation on the project than without. However, hydro-political interactions around the project have not moved from a dispute around sharing water to the discussion of sharing benefits, but rather to discussing both issues simultaneously. An uneven distribution of benefits and different expectations on the project have led the three countries to adopt divergent approaches towards the GERD. Ethiopia and Sudan expect, and are moving jointly to reap, the GERD's benefits accruing from increased power production and power trade for Ethiopia, and expanding irrigated agriculture for Sudan. Given the increased size of the dam and its storage capacity compared to earlier plans and the long negotiations over conducting the required studies on the transboundary impacts of the project, Egypt currently foresees more risks than rewards from the project. Not only is Egypt concerned about the impact of the project on what it considers to be its historical share of the Nile water, but it is also aware of the geopolitical significance of the project, making it vulnerable to Ethiopia's and Sudan's Nile policies, which is perhaps the highest cost of the project.

With the GERD near completion, negotiations over the first filling and operation are expected to face crucial questions that would define the sharing of the benefits and costs of the project. These questions relate to: (1) whether Ethiopia is ready to accept the current uses as a starting point and to adapt the filling and operation to downstream demands defined in terms of these uses; (2) adequate compensation to Egypt for any significant harm; and (3) whether the coordination mechanism would commit the three countries, especially Ethiopia and Sudan, to respect their obligations. These pending questions confirm the suggestion that in cases where potential negative impacts are expected from a water use or hydraulic project, sharing water remains at stake (Dombrowsky 2009).

In light of these contentious issues, strong incentives are needed to induce the three countries to reach compromises. Broadening the ranges of benefits and considering splitting the benefits and costs as suggested by benefit-sharing scholars may be more effective in influencing negotiating positions. This may include sharing the annual operation and maintenance costs of the GERD or future projects or coordination to mobilise international funding for these projects, marketing the power produced by the project and striking deals in other fields. Increasing trade relations, especially virtual water trade imports in livestock and crops from Ethiopia and Sudan to Egypt, and Egyptian investments in the agricultural sector in the two upstream countries have long been proposed, with little progress on the ground.

To achieve progress on these fronts and provide an enabling environment for negotiations, communications between the three countries need not to be confined to the Tripartite National Committee. The regular convening of the tripartite summit proposed by Egypt, and the continuation of the six-party meetings of the ministers of water and irrigation and foreign affairs of the three countries, which have helped transcend the difficulties of talks since their resumption in August 2014, are important to back technical negotiations. A parallel effort to adapt water policies in Egypt and Sudan to the post-GERD era, achieve progress in ongoing attempts to convince Egypt to resume its membership in the NBI, and prepare the public in the three countries to compromises in future negotiations could reduce the intensity of disagreement.

Finally, these negotiations should not be considered as an alternative to a more holistic cooperative process that balances the different uses of riparian states, optimises the development of water resources in the basin, addresses the long-term challenges in the river and fosters regional integration beyond the river.

References

Abdalla, S. and Mohamed, K. (2007). Water Policy of Sudan: National and Co-Basin Approach. Paper presented at the conference 'Water Resources Management in Islamic Countries'. Organised by the Regional Centre on Urban Water Management, Tehran.

Abul Gheit, A. (2014). *Shehadaty: Siyassat Misr Al-kharijiyya 2004–2011* [My Testimony: The Egyptian Foreign Policy 2004–2011], 6th edn. Cairo: Dar Nahdet Masr.

Addis Fortune (2013, 28 April). EEPCo, Chinese Company Ink Billion Dollar Transmission Line Deal. Retrieved from https://addisfortune.net/articles/eepco-chinese-company-ink-billion-dollar-transmission-line-deal/.

Adhanom, T. (2014). The Nile is Symbol of Co-Operation and Collaboration. *Global Dialogue Review*, March–April, 50–57. Retrieved from www.mfa.gov.et/pressMore.php?pg=57.

Agreement (2015). Agreement on Declaration of Principles between the Arab Republic of Egypt, The Federal Democratic Republic of Ethiopia, and the Republic of the Sudan on the Grand Ethiopian Renaissance Dam Project. Signed at Khartoum, Sudan, 23 March 2015.

Ahram (2015, 20 March). The High Committee for Nile Waters Reviews the Declaration of Principles on the GERD. Retrieved from www.ahram.org.eg/NewsQ/370443.aspx.

Ahram Online (2013, 11 June). President Morsi Calls for Unity in Face of Threats to Nile Waters. Retrieved from http://english.ahram.org.eg/News/73683.aspx.

Ahram Online (2016, 5 October). Sisi, Bashir Sign Partnership Agreements, 15 MOUs in High Committee Meetings. Retrieved from http://english.ahram.org.eg/NewsContent/1/64/245252/Egypt/Politics-/Sisi,-Bashir-sign-partnership-agreement,-MoUs-in-h.aspx.

Allan, J.A. (1994). Introduction. In J.A Allan and P. Howell (eds), *The Nile: Sharing a Scarce Resource*. Cambridge: Cambridge University Press.

Allan, J.A. (1999). *The Nile Basin: Evolving Approaches to Nile Waters Management*, Occasional Paper No. 20. London: School of Oriental and African Studies.

Al-Masry Al-Youm (2017, 1 May). Addis Ababa: Egypt Showed Interest in Importing Power from Ethiopia. Retrieved from www.almasryalyoum.com/news/details/1127119.

Al-Monitor (2014, 31 March). Egypt Tries to Woo South Sudan in Nile Water Dispute. Retrieved from www.al-monitor.com/pulse/originals/2014/03/egypt-south-sudan-nile-water-dispute-ethiopia.html#.

Arsano, Y. (2007). *Ethiopia and the Nile: Dilemmas of National and Regional Hydropolitics*. Zurich: Swiss Federal Institute of Technology.

Beyene, A. (2013). Ethiopia's Biggest Dam Oversized. *International Rivers*. Retrieved from www.internationalrivers.org/resources/ethiopia%E2%80%99s-biggest-dam-oversized-expertssay-8082.

Blackmore, D. and Whittington, D. (2008). *Opportunities for Cooperative Water Resources Development on the Eastern Nile: Risks and Rewards*. An independent report of the scoping study team to the Eastern Nile Council of Ministers.

Cascão, A. (2009). Changing Power Relations in the Nile River Basin: Unilateralism vs. Cooperation?. *Water Alternatives*, 2(2), 245–268.

Cascão, A.E. and Nicol, A. (2018). Changing Cooperation Dynamics in the Nile Basin and the Role of the GERD. In Z. Yihdego, A. Rieu-Clarke and A.E. Cascão (eds), *The Grand Ethiopian Renaissance Dam and the Nile Basin: Implications for Transboundary Water Cooperation*. London: Routledge.

Chen, H. and Swain, A. (2014). The Grand Ethiopian Renaissance Dam: Evaluating its Sustainability Standard and Geo-Political Significance. *Energy Development Frontier*, 3(1), 11–19.

Coase, R.H. (1960). The Problem of Social Cost. *The Journal of Law and Economics*, 3, 1–44.

Constitution of the Arab Republic of Egypt (2014). Cairo, adopted on 18 January 2014.

Daily News (2017, 28 January). Al-Sisi Warns Against Strife Instigated by 'Evil People'. Retrieved from https://dailynewsegypt.com/2017/01/28/613127/.

Daily News Egypt (2016, 24 December). Ethiopia Prime Minister accuses Egyptian Institutions of Funding Opposition Groups. Retrieved from https://dailynewsegypt.com/2016/12/24/606458/.

Dombrowsky, I. (2009). Revisiting the Potential for Benefit-Sharing in the Management of Transboundary Rivers. *Water Policy, 11*, 125–140.

Dombrowsky, I. (2010). The Role of Intra-Water Sector Issue Linkage in the Resolution of Transboundary Water Conflicts. *Water International, 35*(2), 132–149.

Dombrowsky, I., Bastian, J., Däschle, D., Heisig, S., Peters, J. and Vosseler, C. (2014). International and Local Benefit Sharing in Hydropower Projects on Shared Rivers: The Ruzizi III and Rusumo Falls Cases. *Water Policy, 16*, 1087–1103.

The East African (2016, 6 February). Egypt Pulls out of Regional Power Pool as it Protests use of Nile Waters. Retrieved from www.theeastafrican.co.ke/news/Egypt-pulls-out-of-power-pool-as-it-protests-use-of-Nile-waters/2558-3065704-x4qbu2z/index.html.

The East African (2017, 2 February). President Kiir Dismisses Claims of Juba–Egypt Conspiracy. Retrieved from www.theeastafrican.co.ke/news/Kiir-dismisses-claims-of-Juba-Egypt-conspiracy/2558-3796954-y99e61z/index.html.

Ethiopian News Agency (2015, 24 November). Ethiopia, Sudan Discuss Economic Integration. Retrieved from www.ebc.et/web/ennews/-/ethiopia-sudan-discuss-economic-integration.

Ethiopian News Agency (2017, 30 April). Ethiopia Finalizes Feasibility Study to Export Electricity to Egypt. Retrieved from www.ena.gov.et/en/index.php/economy/item/3129-ethiopia-finalizes-feasibility-study-to-export-electricity-to-egypt.

European Union (2016). Press Release: The European Union Supports Egypt's National Water Resources Plan. Brussels: EU External Action Service.

Fahmy, N. (2014). Ethiopia's Dam is a Major Risk to Egypt's Water Security. *Global Dialogue Review*. Retrieved from www.globaldialoguereview.com/forum/ethiopias-dam-is-a-major-risk-for-egypts-water-security-nabil-fahmy/.

Gebremedhin, K. (2015). The Khartoum Declaration on GERD and the 1997 Convention on the Law of Non-Navigational Uses of International Water Courses. *The Ethiopian Observatory*. Retrieved from http://ethiopiaobservatory.com/2015/04/01/the-khartoum-declaration-on-gerd-the-1997-un-convention-on-the-law-of-non-navigational-uses-of-international-watercourses/.

Guariso, G. and Whittington, D. (1987). Implications for Ethiopian Water Development on Egypt and Sudan. *International Journal of Water Resources Development, 3*(2), 105–114.

Hammond, M. (2013). *The Grand Ethiopian Renaissance Dam and the Blue Nile: implications for Transboundary Water Governance*, Discussion Paper No. 1307. Canberra: Global Water Forum.

Hensengerth, O., Dombrowsky, I. and Scheumann, W. (2012). *Benefit-Sharing in Dam Projects on Shared River*, Discussion Paper No. 6. Bonn: German Development Institute.

Horn Affairs (2014, April 8). Ethiopia Rejects Egypt's Offer to Finance the Renaissance Dam. Retrieved from http://hornaffairs.com/en/2014/04/08/ethiopia-reject-egypt-finance-renaissance-dam/.

International Monetary Fund (2014). The Federal Democratic Republic of Ethiopia. Country Report No. 14/303. Washington, DC: Author.

International Panel of Experts (IPoE) (2013). Grand Ethiopian Renaissance Dam: Final Report.

International Rivers (2012). *Field Visit Report: GERD Project*. Berkeley, CA: Author.

Kahsay, T.N., Kuik, O., Brouwer, R., and Van der Zaag, P. (2015). Estimations of the Transboundary Economic Impacts of the Grand Ethiopian Renaissance Dam: A

Computable General Equilibrium Analysis. *Water Resources and Economics*. Doi: http://dx.doi.org/10.1016/j.wre.2015.02.003.

Kahsay, T.N., Kuik, O., Brouwer, R. and Van der Zaag, P. (2018). Economic Impact Assessment of the Grand Ethiopian Renaissance Dam under Different Climate and Hydrological Conditions. In Z. Yihdego, A. Rieu-Clarke and A.E. Cascão (eds), *The Grand Ethiopian Renaissance Dam and the Nile Basin: Implications for Transboundary Water Cooperation*. London: Routledge.

King, A. (2013). *An Assessment of Reservoir Filling Policies under Changing Climate for Ethiopia's Grand Renaissance Dam*. Unpublished Master's Thesis. Drexel University, Pennsylvania.

McCartney, M. and Girma, M. (2012). Evaluating the Downstream Implications of Planned Water Resources Development in the Ethiopian Portion of the Blue Nile. *Water International, 37*(4): 362–379.

Massachusetts Institute of Technology (MIT) (2014). *The Grand Ethiopian Renaissance Dam: An Opportunity for Collaboration and Shared Benefits in the Eastern Nile Basin. A Brief to Riparian Nations of Ethiopia, Sudan and Egypt*. Cambridge: Author.

Massachusetts Institute of Technology (MIT) (2015). Comments on the AMICUS brief report for the workshop titled '*The Grand Ethiopian Renaissance Dam: An Opportunity for Collaboration and Shared Benefits in the Eastern Nile Basin*' by Panel of Egyptian experts assembled by the Ministry of Water and Irrigation of the Arab Republic of Egypt. Cambridge: Author.

Matthews, N., Nicol, A. and Seide, W.M. (2013). Constructing A New Water Future?: An Analysis of Ethiopia's Current Hydropower Development. In T. Allan, M. Keulertz, S. Sojamo and J. Warner (eds), *Handbook of Land and Water Grabs in Africa*. London and New York: Routledge.

Ministry of Finance and Economic Development of the Federal Democratic Republic of Ethiopia (2010). Growth and Transformation Plan 2010/11–2014/15. Addis Ababa: Author.

Ministry of Foreign Affairs of the Arab Republic of Egypt (2015a). Spokesperson's Office. Media follow-up. Cairo: Author.

Ministry of Foreign Affairs of the Arab Republic of Egypt (2015b). Spokesperson's Office. Nile Historical Treaties Untouchable. Cairo: Author.

Ministry of Foreign Affairs of the Arab Republic of Egypt (2014a). *Fahmy Delivers a Lecture at the Royal Institute for International Development*. Cairo: Author.

Ministry of Foreign Affairs of the Arab Republic of Egypt (2014b). Egypt's Perspective Towards the Ethiopian Grand Renaissance Dam Project (GERDP). Cairo: Author.

Ministry of Foreign Affairs of the Arab Republic of Egypt (2014c). No New Policy for Egypt Towards the Renaissance Dam. Cairo: Author.

Ministry of Foreign Affairs of the Federal Democratic Republic of Ethiopia (2012). Misguided Fear or Deliberate Distortions in the Nile Basin Again?. Addis Ababa: Author.

Ministry of Foreign Affairs of the Federal Democratic Republic of Ethiopia (2016). The Africa 2016 Business Forum in Sharm el-Sheikh. Addis Ababa: Author.

Ministry of Foreign Affairs of the Republic of Sudan (2014a). Agreement signed for establishing joint Sudanese–Ethiopian forces. Khartoum: Author.

Ministry of Foreign Affairs of the Republic of Sudan (2014b). The Renaissance Dam. Retrieved from http://mofa.gov.sd/new/en/more.php?main_id=2&sub_id=18&id=540.

Ministry of Information of the Federal Democratic Republic of Ethiopia (2002). Foreign Affairs and National Security Policy Strategy. Addis Ababa: Author.

Ministry of Water Resources and Irrigation of the Arab Republic of Egypt (2016). The Phases of the One Million Feddan Project. Retrieved from www.mwri.gov.eg/Million-Level.aspx.

Ministry of Water Resources of the Federal Democratic Republic of Ethiopia (2001a). Ethiopian Water Sector Strategy. Addis Ababa: Author.

Ministry of Water Resources of the Federal Democratic Republic of Ethiopia (2007). Ethiopia's Hydropower Projects. Retrieved from www.mowr.gov.et/index.php?pagenum=4.3&pagehgt=1000px.

Mohamed, K. (2015). Ma'akhzwathiqati'lan al-mabade' hawla sad elnahda men al-montalaq al-ilmy al-handasy [Reservations on the Declaration of Principles on GERD from a Technical Point of View]. *Sudacon Net*, Retrieved from www.sudacon.net/2015/04/blog-post_6.html.

Mulat, A.G. and Moges, S.A. (2014). Assessment of the Impact of the Grand Ethiopian Renaissance Dam on the Performance of the High Aswan Dam. *Journal of Water Resources and Protection*, 6, 583–598.

Negash, M., Hassan, S., Muchie, M. and Moges, A.G. (2015). Perspectives on the Declaration of Principles Regarding the Grand Ethiopian Renaissance Dam. *The Thinker*, 65, 56–61.

The New Arab (2015, 24 November). Sudan Files UN Complaint Against Egypt. Retrieved from www.alaraby.co.uk/english/news/2015/11/24/sudan-files-un-complaint-against-egypt.

Nile Basin Initiative (2012). *State of the River Nile Basin*. Entebbe: Author.

Nour Eddin (2014). *Misrwa Dewal Manabe' el-Nil: al-hayah, walmeyah, walsodoud walsera'* [Egypt and Upstream Nile Countries: Life, Water, Dams and Conflict]. Cairo: Dar Nahdet Masr.

Nour Eddin (2015). Mofawadat sad el-nahdaakbarhazima li Misrwaotaleb al-ra'es be tadweel al-qadiyya [The Negotiations over the GERD is the Biggest Blow to Egypt and I Call Upon the President to Internationalise the Issue], interview with Journal Misr, 29 November.

Raslan, H. (2013). Ro'yanaqdiyya li idarat sad el-nahda [A Critical View of Egypt's Management of the Renaissance Dam Crisis]. *Al-Siyassa Al-Dawliyya [Journal of International Relations]*, 199.

The Reporter (2014, 29 March). Ethiopia Admonishes Eritrea over South Sudan's Conflict. Retrieved from http://allafrica.com/stories/201403311126.html.

The Reporter (2016, 29 October). Ethiopia Closely Watches as Egypt bids to Join Peacekeeping in South Sudan. Retrieved from http://thereporterethiopia.com/content/ethiopia-closely-watches-egypt-bids-join-peacekeeping-south-sudan.

Sadoff, C. and Grey, D. (2002). Beyond the River: The Benefits of Cooperation on International Rivers. *Water Policy*, 4, 389–403.

Sadoff, C. and Grey, D. (2005). Co-Operation on International Rivers: A Continuum for Securing and Sharing Benefits. *Water International*, 30(4), 1–8.

Sadoff, C. and Grey, D. (2008). Why Share: The Benefits (and Costs) of Transboundary Water Management. In C. Sadoff, T. Grieber, M. Smith, and G. Bergkamp (eds), *Share: Managing Water across Boundaries*. Gland: International Union for Conservation of Nature.

Scheumann, W., Dombrowsky, I. and Hensengerth, O. (2014). Dams on Shared Rivers: The Concept of Benefit-Sharing. In A. Bhaduri, J. Bogardi, J. Leentvaar, S. Marx (eds), *The Global Water System in the Anthropocene: Challenges for Science and Governance*. Berlin: Springer.

Sharaqi, A. (2011). The Impact of the Ethiopian Dam Projects on the Nile Water. Paper presented to the Annual Conference of the Institute for Africa Research and Studies, Cairo University, Giza.

State Information Service of the Arab Republic of Egypt (2015). Joint Communiqué on Occasion of President Sisi's visit to Ethiopia. Cairo: Author.

Strategic Thinking of East Africa (2015, 17 October). Ethiopia's Grand Renaissance Dam Will Generate 750 MW Soon. Retrieved from www.strathink.net/ethiopia/5294/.

Sudan News Agency (2013, 7 October). Ethiopia calls on Sudan and Egypt to Contribute to Establishment of Renaissance Dam. Retrieved from http://allafrica.com/stories/201310080341.html.

Sudan Tribune (2012, 8 March). Sudan's Bashir Supports Ethiopia's Nile Dam Project. Retrieved from www.sudantribune.com/spip.php?article41839.

Sudan Tribune (2016, 28 October). Sudan to Build Power Transmission Line from Ethiopia's GERD: Minister. Retrieved from www.sudantribune.com/spip.php?article60680.

Swain, A. (2002). The Nile River Basin Initiative: Too Many Cooks, Too Little Broth. *SAIS Review, 22*(2), 293–308.

Tawfik, R. (2015). *Revisiting Hydro-Hegemony from a Benefit-Sharing Perspective: The Case of the Grand Ethiopian Renaissance Dam*, Discussion Paper No. 5/2015. Bonn: German Development Institute.

Tawfik, R. (2016). Changing Hydro-Political Relations in the Nile Basin: A Protracted Transition. *The International Spectator, 51*(3), 67–81.

Tvedt, T. (2010). Some Conceptual Issues Regarding the Study of Inter-State Relationships in the Nile Basin. In T. Tvedt (ed.), *The River Nile in the Post-Colonial Age; Conflict and Cooperation Among the Nile Basin Countries* (pp. 237–246). London: I.B.Tauris.

United Nations, Convention on the Law of the Non-Navigational Uses of International Water Courses, Adopted on 21 May 1997.

Veilleux, J. (2013). The Human Security Dimensions of Dam Development: The Grand Ethiopian Renaissance Dam. *Global Dialogue, 15*(2), 1–15.

Waterbury, J. (2002). *The Nile Basin: National Determinants of Collective Action.* New Haven, CT: Yale University Press.

Wheeler, K.G. (2018). 'Managing Risks While Filling the Grand Ethiopian Renaissance Dam. In Z. Yihdego, A. Rieu-Clarke and A.E. Cascão (eds), *The Grand Ethiopian Renaissance Dam and the Nile Basin: Implications for Transboundary Water Cooperation.* London: Routledge.

Whittington, D., Waterbury, J. and Jeuland, M. (2014). The Grand Renaissance Dam and Prospects for Cooperation on the Eastern Nile. *Water Policy, 16*: 595–608.

World Bulletin (2014, 19 April). South Sudan's Kiir Reassures Ethiopia PM on Dam, Egypt Military Accord. Retrieved from www.zehabesha.com/s-sudans-kiir-reassures-ethiopia-pm-on-dam-egypt-military-accord/.

Zeitoun, M. and Warner, J. (2006). Hydro-Hegemony: A Framework for Analysis of Transboundary Water Conflicts. *Water Policy, 8*: 435–460.

Zenawi, M. (2011). Speech on Launching the GERD. Retrieved from www.meleszenawi.com/ethiopian-pm-meles-zenawi-speech-on-launching-gerd-text-and-videos/.

7 Analysing the economy-wide impacts on Egypt of alternative GERD filling policies

Brent Boehlert, Kenneth M. Strzepek and Sherman Robinson

Introduction

Ethiopia is currently constructing the Grand Ethiopian Renaissance Dam (GERD) on the Blue Nile, one of two primary tributaries of the Nile River and the primary source of water supply in downstream Egypt. The economic and social benefits of this dam to Ethiopia are likely to be quite significant: the GERD will generate tremendous amounts of hydropower, thereby potentially improving regional energy security and livelihoods. Egypt has expressed concern over the GERD, noting that the filling of the GERD reservoir would likely reduce the reliability of Nile River flows, with potentially grave impacts on the economy of Egypt. Despite these concerns, the extent to which Egypt's economy will be affected by GERD filling has not been evaluated.

This study couples three models to analyse how GERD filling combined with variable Lake Nasser inflows will impact Egypt's hydropower generation and irrigation deliveries over a future twenty-year period, and how these will affect economy-wide indicators including GDP. Because future Nile flows are highly uncertain, the study takes a risk-based approach: for each of nine GERD filling policies, 100 seventeen-year time series of Lake Nasser inflows are generated from a water systems model of the Nile. These are routed into a constrained non-linear programming model of the Egyptian water system that is coupled dynamically with the International Food Policy Research Institute's (IFPRI's) computable general equilibrium (CGE) model of Egypt's economy. This dynamic model is then used to evaluate the potential effects of GERD filling policies on the economy of Egypt.

Water system models and CGE models have been applied independently to Egypt and the Nile Basin in many previous studies. Strzepek *et al.* (2007), for example, employ the Egypt CGE model in a study of the value of the High Aswan Dam (HAD) to the Egyptian economy, and conclude that the dam increased annual 1997 GDP by between 2.7 per cent and 4.0 per cent. Robinson *et al.* (2008) use the same CGE model to evaluate the indirect impacts of the HAD on Egypt's economy, and Robinson and Gehlhar (1996) used an earlier CGE model of Egypt to estimate the shadow price of water and analyse the implications of water shortages on the Egyptian economy.

Other research has evaluated the potential benefits of cooperation over Nile water resources. Wheeler *et al.* (2016) use a water systems model to evaluate the effects of GERD filling policies on energy generation and water supply reliability in Egypt, and find that risks to Egypt can be limited through minimum GERD releases during filling and a HAD drought management policy. Our study takes a similar approach to Wheeler *et al.*, but uses a CGE model to evaluate the economic implications of GERD filling policies and agreed releases. Whittington *et al.* (2005) apply a constrained NLP model of the Nile Basin to evaluate the potential economic benefits of cooperation over water in the Nile River system. They find that cooperation could generate between $7 billion and $11 billion in annual benefits, not accounting for infrastructure costs. Similar to this study, the objective function of their water management model maximises the sum of irrigation and hydropower revenues; however, it does so across the entire Nile Basin rather than for only Egypt and does not allow for substitution and interaction with world petroleum and food markets, as does a CGE model.

Block, Strzepek and Rajagopalan (2007) apply an NLP model to the Ethiopian portion of the Nile River Basin. Their model takes as inputs 30- to 100-year runoff and reservoir evaporation values to generate hydropower and irrigation revenues as outputs. Like this study, Block, Strzepek and Rajagopalan aim to evaluate potential economic outcomes under various climate futures. However, their decision variables centre on management of reservoirs in Ethiopia rather than Egypt. Jeuland (2010) uses Monte Carlo simulation with a hydroeconomic modelling framework to evaluate two potential sizes of an Ethiopian hydropower project on the Blue Nile, and in so doing, illustrate the advantages of incorporating economic considerations and uncertainty into a water planning model.

While this past research has either evaluated water issues using CGE or economic issues using a water-planning model, no prior studies in the Nile Basin have linked a water-planning model with a CGE to form a dynamic water system-economy modelling framework. Robinson and Gueneau (2013) have created the first such linked framework through the CGE-W model platform, which couples a Pakistan CGE model with the Regional Water System Model for Pakistan (RWSM-Pak) – adapted from the World Bank's Indus Basin Model Revised (IBMR) – that has updated routing, water demand and water stress components. We adopt the same CGE-W modelling framework in this study by linking a CGE model of Egypt with a water system model that also includes hydropower generation as an output. Our modelling approach captures a wide range of potential water availability outcomes for Egypt and the resulting impacts to Egypt's economy. The model operates with perfect knowledge of weather and other conditions during the current year, but has no knowledge of weather in future years.

In the following sections, we provide a background on the Nile River Basin, explain our modelling approach, summarise the results of the analysis and conclude by discussing our findings and paths for further research.

The Nile River Basin: geographic and institutional context

The Nile River and its tributaries flow 4,130 miles through eleven countries along a path from the region of Lake Victoria in Uganda to the Mediterranean Sea (see Figure 10.1 from Wheeler, Chapter 10). The Nile's two primary tributaries are the White Nile and the Blue Nile; the White Nile has a much larger drainage area and contributes relatively steady flow over the year, whereas the Blue Nile contributes significantly more flow overall, but over a much shorter period of the year. Of average historical Nile inflow to Egypt, 59 per cent comes from the Blue Nile, 13 per cent from the Atbara (a tributary that enters the Nile downstream of the confluence of the White Nile and Blue Nile) and 28 per cent from the White Nile.

This paper focuses on Ethiopia, which encompasses most of the Blue Nile Basin, and Egypt, the furthest downstream country in the system. The HAD, the focus of the Egyptian water system model used in this study, is located on the Nile at the border between Egypt and Sudan. Water use is highest in Egypt, largely due to the nation's extensive irrigation system, industrial development and large population. Water use is lowest in Ethiopia, which has predominantly subsistence rain-fed agriculture and rural populations. Through the Nile Basin treaty, enacted in 1959, Egypt and Sudan unilaterally allocated the agreed-upon mean annual Nile flow of 84 billion cubic metres (BCM). Ten BCM was allocated to evaporation from reservoirs, leaving 55.5 BCM to Egypt and 18.5 BCM to Sudan. Thus, the 1959 Nile Basin treaty allocated 75 per cent of the mean annual Nile Flow to Egypt, 25 per cent to Sudan and none to Ethiopia (for more details, see Howell *et al.* 1995).

Ethiopia has a clear reason for its interest in constructing reservoirs on the Blue Nile system: the steep terrain in Ethiopia provides opportunities for much hydropower generation. To shed light on the possible impacts of Ethiopian water resources development on Egypt's water and economic security, this paper examines the short-term impacts of GERD filling policies on the Egyptian economy using a risk-based approach. Wheeler *et al.* (2016) also use a risk-based approach to evaluate the impacts of GERD filling policies on water supply reliability to Egypt and find that reliability could fall under some scenarios. The current study extends that work by evaluating the effect on Egypt's economy.

Modelling methodology

Our modelling methodology is designed to evaluate the impact of GERD filling policies on Egypt's economy. Numerous synthetic hydrological time series (from Zhang *et al.* 2016) are routed through a Nile water systems model under several GERD fill policies, and the resulting Lake Nasser inflows are inputs to the dynamically coupled water systems and economy-wide model of Egypt (Figure 7.1). This modelling approach accounts for the inherently dynamic interactions between water and economic systems: water availability and management affect the distribution of water, which affects crop yields and hydropower generation.

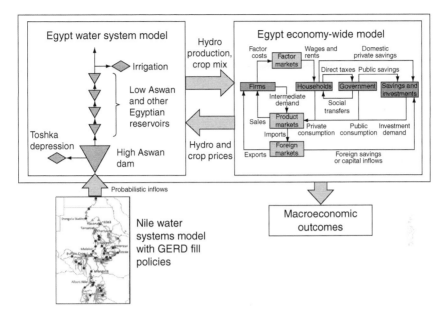

Figure 7.1 Modelling framework.

These physical outcomes affect agricultural and energy supplies, and the economy then reacts by reallocating production factors through market mechanisms and price changes. These changes in prices affect water managers' decisions on the timing of hydropower generation and on farmers' cropping decisions in the following year. In this section, we describe this modelling framework shown in Figure 7.1, including the stochastic Lake Nasser inflows used for this analysis, the Egypt water system model, the IFPRI CGE model of Egypt's economy and the linkage between the water and CGE models.

Lake Nasser inflows

The inflows to Lake Nasser are driven by nine GERD filling policies, which are made up of combinations of three assumed GERD fill duration scenarios and three assumed GERD release requirement scenarios. The fill duration scenarios are the minimum period over which the GERD can be filled to the top of conservation pool, and include unconstrained filling, three-years to fill and ten-years to fill. Under the ten-year fill duration scenario, top of conservation cannot be reached until year ten of filling, whereas under the unconstrained scenario, the GERD can be filled as soon as sufficient Blue Nile flow is available. The scenarios of required releases from the GERD include no minimum, 15 BCM, or 30 BCM annually. To capture the uncertainty of Nile inflows that will occur during the GERD filling and thus the uncertain inflows to Lake

Nasser over the filling period, 100 stochastic Nile streamflow time series were generated using wavelet analysis (see Zhang *et al.* 2016). These flows were routed through a Mike Hydro water system model of the Nile above Lake Nasser, with no GERD, and with GERD in place and operated under the nine filling scenarios. The model simulation assumed a HAD starting elevation of 175 m-asl. The final output of the water system model for each of the ten policy scenarios was an ensemble of 100 members of seventeen-year time series of Lake Nasser inflows.

The nine fill policies produce Lake Nasser inflows that align with expectations, with the unconstrained duration scenarios producing the largest initial reductions, and the no minimum GERD release requirement scenarios producing more variable time series (Figure 7.2). In the case of the unconstrained fill duration, no release requirement scenario, the 25th to 75th percentile (interquartile range) of annual Lake Nasser inflows in year one fall by between 20 BCM and 35 BCM relative to the no GERD average inflow. Under this

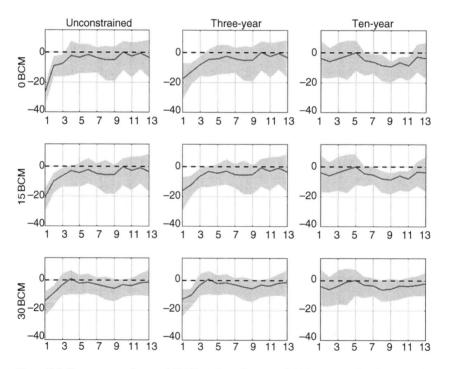

Figure 7.2 Divergence of annual HAD inflows from no GERD average levels across the three required fill duration and three release scenarios.

Source: own analysis.

Notes
Horizontal axis is years, with GERD fill starting at year 1. Vertical axis is annual HAD inflows in BCM. Black line is the mean across the 100 stochastic inflows and the grey shaded region is the inter-quartile range.

scenario, at the 5 per cent risk level (or the fifth worst outcome out of 100 runs) year one inflow falls by 40 BCM, or by over half the average inflows without the GERD. Note that under the ten-year fill duration, 30 BCM required release scenario, the GERD is often unable to fill within ten years.

The Egypt water system model

The stochastic outflows from the Mike Hydro model are routed into a stylised water system model of Egypt, which simulates the operation of the HAD, with releases for the generation of hydroelectricity supplies and water supplies to irrigation and other sectoral uses in the Nile valley in Egypt. The Egypt water system model is an optimisation model that balances monthly upstream inflows against hydropower releases, Toshka spills, net evaporation, seepage and changes in reservoir volume. The three primary components of the model are inflows, the HAD itself and downstream demands (upper left section of Figure 7.1). The HAD component of the model encompasses the objective function, all operating rules, hydropower generation, reservoir evaporation and seepage, and physical characteristics of the reservoir such as elevation–volume and elevation–surface area relationships. Lastly, the downstream system that is modelled includes Egyptian hydropower facilities and irrigation. In the objective function, downstream irrigation demands, maintenance of reservoir elevations and the Aswan complex hydroelectricity generation are the drivers of dam releases.

We formulated the above, simplified representation of the Egyptian Water System into a constrained non-linear programming model. The model has a single objective function, a continuity equation to ensure mass balance within the reservoir, and multiple operational constraints that define the hydrology, agriculture and hydropower of the system. The model was constructed using the Generalized Algebraic Modelling System (GAMS) software package.

The release rules of the HAD first ensure that downstream demands are met and then maximise hydropower production as a secondary goal. As a result, the model's objective function maximises weighted irrigation revenues plus hydropower generation, and subtracts reservoir spills to ensure they only occur when necessary. The objective function is as follows:

$$Maximize \ Z = w_a \sum_c^{n^c} p_c^c L_c x_c + \sum_t^{12} \gamma_{1t} - \sum_t^{12} (R_t^T + R_t^S)$$

Where:
Z = value of the objective function
w_a = constant multiplier on agricultural revenue, set at 10
p_c^c = price per crop (updated annually from the CGE model)
x_c = yield per crop
L_c = land allocated to each crop
n^c = number of crops

γ_{1t} = hydropower generated each month at HAD
R_t^T = releases through the Toshka canal
R_t^S = releases not used for hydropower generation (i.e. spills)

The monthly continuity equation on the High Aswan Reservoir states that accumulation is equal to inflow minus releases, demands, evaporation and seepage.

$$V_t - V_{t-1} = Q_t - R_t^H - R_t^S - R_t^T - Evap_t - Seep_t; \ \forall t$$

Where the left-hand side is the change in the reservoir volume from month $t-1$ to month t, Q_t is the inflow to the reservoir during month t (from Mike Hydro). R_t^H and R_t^S are the modelled releases from the reservoir through turbines and spills, R_t^T are the modelled releases to the Toshka canal and pumping for irrigation west of the Reservoir, $Evap_t$ are the monthly evaporation volumes from (from Allen *et al.* 2002), and $Seep_t$ are the monthly seepages that occur (from Mobasher 2010).

There are also several reservoir operations constraints. First, monthly reservoir elevations (h_t) must be between the minimum and maximum allowable elevations.

$$h^{min} \le h_t \le h^{max}$$

The reservoir elevation must be less than or equal to 175 m-asl at the end of each Egyptian water year (31 July). This elevation provides enough flood storage to ensure that the reservoir is capable of storing peak season flows, but also is high enough to provide the storage needed to protect against extended drought.

$$h_{12} \le 175$$

Total monthly releases from the HAD must not exceed the maximum level allowable through the turbines and spillway.

$$R_t^H + R_t^S \le R^{max}$$

To meet downstream demands, annual releases as spill and through the turbines must sum to 55.5 BCM, unless: (a) additional releases are required to meet the 31 July elevation requirement above; or (b) reservoir elevations drop below a certain threshold and the release requirement is lowered. The additional 1 August releases requires separating R_t^S into two components: $R_t^S = R_t^{S1} + R_t^{S2}$, where R_t^{S1} is spill in excess of turbine capacity to meet the annual minimum release requirement, and R_t^{S2} is any additional spill needed to reach the 31 July elevation requirement. Note that both forms of spill are penalised in the objective function, so will be minimised. The annual release requirement is then formulated as follows.

$$\sum_t^{12} (R_t^H + R_t^{S1}) = R^{REQ}$$

R^{REQ} is 55.5 BCM when annual average reservoir elevations exceed 154 m-asl, and 52.5 BCM when elevations are below this level. This elevation threshold is based on conditions in 1988, when mean annual reservoir elevations reached 154 m-asl and 52.5 BCM was released (Mobasher 2010).

The surface area and volume of the HAD reservoir (used for evaporation calculations) are exponential functions of water elevation (from Mobasher 2010).

$$A_t = 1166.8 e^{0.0391(h_t - 137.5)}$$

$$V_t = 18.404 e^{0.0494(h_t - 137.5)}$$

Monthly seepage is a fixed seepage loss (ω; from Mobasher 2010) multiplied by days in the month (α_t).

$$Seep_t = \alpha_t \omega$$

Monthly reservoir evaporation is the surface area of the reservoir (A_t) times the potential evapotranspiration (PET) rate.

$$Evap_t = A_t PET_t$$

The Toshka canal also has a set of constraints that dictate when water is spilled from the HAD into the canal. When reservoir elevations exceed 178 m-asl, water is spilled from the Toshka canal, which is a product of days per month (α_t), a constant, and a power function of head over 178 m-asl (from Mobasher 2010). This Toshka releases term also includes pumping for the Toshka irrigation project.

$$R_t^T = 0.019 \alpha_t \, (h_t - 178)^{1.667} + pump_t; \, h_t > 178$$

A set of constraints also governs land use and irrigation. First, land use per crop (L_c) during a year must not exceed land available for each crop (L_c^{avail}; updated annually from the CGE model).

$$L_c \leq L_c^{avail}$$

Irrigation each month and for each crop (I_{tc}) must equal land use times monthly and per-crop evapotranspiration (ET_{tc}; from Strzepek *et al.* 2007), divided by basin-level irrigation efficiency (IE) in Egypt (from Keller and Keller 1995).

$$I_{tc} = L_c \frac{ET_{tc}}{IE}$$

Finally, total irrigation use across crops in a given month must not exceed the total releases from the reservoir.

$$\sum_c^{n^c} I_{tc} \leq R_t^H + R_t^S$$

Irrigation revenues (Y^I) are then equal to the sum over crops of crop price (p_c^c) times land used per crop times yield per crop (x_c; base yield from FAO PriceStat, price is updated annually from the CGE model). This formulation assumes that farmers have perfect foresight of water deliveries for the year, which is reasonable for a system regulated by a set of reservoirs with a collective storage equivalent to multiple years of downstream supply. For crops with annual yield response factors (FAO 1998) of one or greater, farmers faced with a water deficit are assumed to irrigate proportionally less land. For crops with yield response factors of less than one, we assume farmers deficit irrigate the land to obtain higher production levels for a given amount of water.

$$Y^I = \sum_c^{n^c} p_c^c L_c x_c$$

There are also several constraints that define hydropower generation. Monthly reservoir releases through the turbines at each facility h (R_{ht}^H) must be between the minimum and maximum turbine flow (R_h^{HMin} and R_{ht}^{HMax} from Mobasher 2010; flow outside of these boundaries is spilled).

$$R_h^{HMin} \leq R_{ht}^H \leq R_{ht}^{HMax}$$

At HAD, maximum turbine flow declines with the square root of head (h_t minus 110 m-asl, the average water table below the dam), which is multiplied by an coefficient (σ) calculated such that the right-hand side of the equation produces maximum turbine capacity at design head, and number of days per month (α_t).

$$R_{1t}^{HMax} = \sigma \alpha_t \sqrt{head_{1t}}$$

Hydropower generated each month and at each facility (γ_{ht}) is equal to a dam-specific coefficient (δ_h) times head that month ($head_{ht}$) times mean monthly flow (converted to cubic metres per second, or cms) times turbine efficiency (ε_h; all from Mobasher 2010).

$$\gamma_{ht} = \delta_h head_{ht} R_{ht}^H \varepsilon_h$$

Hydropower revenues in a month (Y_{ht}^H) for each facility are equal to the energy (converted from power by multiplying by hours in a month β_t) times the energy price (p^h; updated annually from the CGE).

$$Y_{ht}^H = \beta_t \gamma_{ht} p^h$$

To ensure that the Egypt water system NLP is properly modelling reservoir dynamics, including potentiometric head for hydropower generation, the model was 'validated' by fixing monthly releases at historical levels and then comparing the modelled reservoir elevations to the observed record. We obtained data

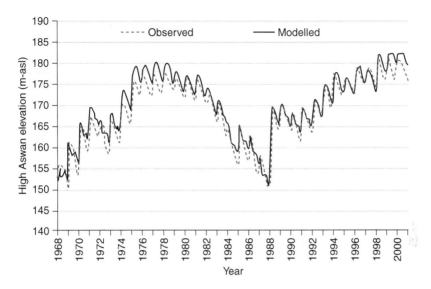

Figure 7.3 Comparison of modelled and observed High Aswan reservoir elevations, 1968 to 2001.

Source: own analysis, data from Mobasher 2010.

on releases and reservoir elevations for the validation period of 1968 to 2001 from Mobasher (2010). Inflows to the reservoir were set at the observed levels for the validation period, and all modelled reservoir relationships (e.g. volume to elevation, evaporation and seepage) were maintained. Modelled and observed reservoir elevations are compared in Figure 7.3. Overall, modelled elevations match their observed counterparts well.

The Computable General Equilibrium model of Egypt

CGE models provide a full accounting of production, consumption, and trade in a particular economy. They have become widely used in policy analysis in developing countries since their first applications in the mid-1970s. In this analysis, we apply a modified version of IFPRI's Standard CGE model, which is written in GAMS. Below, we provide a brief description of the Egypt CGE model. For a more detailed discussion of the IFPRI standard model, see Lofgren *et al.* (2001). For more detail on the Egypt application of the standard model, see Strzepek *et al.* (2007) and Robinson *et al.* (2008).

The Egypt CGE model follows the disaggregation of a Social Accounting Matrix (SAM), and was written as a set of simultaneous equations, many of which are non-linear. These equations define different actors' behaviour, which in part follows simple rules captured by fixed coefficients (e.g. ad valorem tax rates). The model captures production and consumption behaviour through non-linear, first-order optimality conditions of profit and utility maximisation. The equations also include a set of 'system constraints' that define macroeconomic equilibria (balances for savings-investment, the government, and current-account of the rest of the world) and equilibrium in markets for factors and commodities.

The upper right panel of Figure 7.1 provides an overview of the links between the components of the standard IFPRI CGE model employed here, where the arrows represent payment flows. Disaggregation of the SAM determines the disaggregation of representative households, factors and commodities. The model includes 'real' flows for commodities or factor services that have arrows in the opposite direction – with the exception of taxes, transfers and savings. The activities carry out production and allocate their income from output sales to intermediate inputs and factors.

Producers are assumed to maximise profits subject to prices and a nested technology in two levels. At the top, output is a Leontief function of aggregates of value-added and intermediate inputs. At the second level, aggregate value-added is a constant elasticity of substitution (CES) function of factors, whereas the aggregate intermediate input is a Leontief function of disaggregated intermediate inputs. The agricultural sector is disaggregated to better represent the impact of droughts and water shortfalls. The water systems model provides impacts to the value added by imposing a reduction on agricultural sector production due to water stress as well as impacts on hydroelectric production based on changes in releases from the HAD. This is described further in Robinson and Gueneau (2013).

The model has two seasons and distinguishes summer land (for cotton, rice, maize, sorghum and other summer crops) from winter land (for wheat, legumes, long and short berseem, and other winter crops). With the exception of vegetables and sugar cane, which are included annually, the various crops use land in only one season. Each year, land is allocated efficiently across crops according to profitability.

Producers take prices as given when making their decisions, based on the assumption that they are small relative to the market and have no perfect forecast. After meeting home consumption demands, the model allocates outputs between the domestic market and exports in shares that respond to changes in the ratio between domestic and world producer prices. Supplies of exports in world markets follow the small-country assumption: they are absorbed by infinitely elastic demands at fixed prices. Supplies from domestic and world producers meet domestic market demands. For all commodities, the ratio of imports to domestic output demand responds to changes in the relative prices of imports and domestic output sold domestically. An infinitely elastic supply of imports at fixed prices allows for meeting import demand. In domestic markets for domestically sourced products, quantities demanded and supplied are assured to be equal through flexible prices.

Producers' factor costs are passed on as receipts to the household block in shares that reflect endowments. The household block may also receive transfers from other households, the government (which are CPI-indexed) and the rest of the world (fixed in foreign currency). Households spend these incomes on savings, direct taxes, transfers to other institutions and consumption. We model savings, direct taxes and transfers as fixed income shares. For both home-consumed and market-purchased goods, consumption is divvied across commodities according to LES (Linear-Expenditure-System) demand functions, which are derived from utility maximisation.

The government receives taxes from households and transfers from the rest of the world, which it then spends on consumption, transfers to households and savings. The current account of the balance of payments (i.e. the rest of the world) receives foreign currency from exports, and then spends these earnings on imports, transfers to government and on foreign savings. Finally, savings from all institutions are collected in the investment account and used to finance domestic investment.

Each model solution provides a wide range of economic indicators (e.g. GDP; consumption and incomes for representative households; sectoral production and trade volumes; factor employment; commodity prices; and factor wages). We use changes in GDP to approximate changes in total welfare (consumer plus producer surplus), and changes in rural and urban incomes to evaluate distributional effects.

Linking the water and CGE models

The CGE and water models are linked in a two-step solution approach for each year. First, the CGE model is solved at the 'beginning' of the year based on projected factor supplies, productivity trends, world prices and water supplies. This solution determines the cropping pattern chosen by farmers. The physical cropped area is calculated based on the land allocation in the CGE model. This represents changes in cropping patterns due to market price fluctuations, and is sent to the water model to determine the demand for water in agriculture.

The water model is run using this cropping pattern to determine the availability of water for agriculture by crop in the current year. The water model run determines reservoir releases to maximise irrigated agriculture profits and hydropower generation. The model optimises agricultural profits by allocating water to the specified crop areas to determine crop yields that take advantage of the non-linear diminishing returns to scale water production functions. These optimised yields are then used as inputs for the second run of the CGE model. For each crop, the reduced production is assumed to be the proportion of yield lost due to water stress. This production loss is directly applied as a reduction in total factor of production (TFP) of the concerned crop in the CGE model. Impacts to energy production occur because of changes in hydropower production from the HAD due to reduced reservoir levels and/or turbine releases. These losses affect the electricity sector in the CGE model. This loss is measured as a ratio of the hydropower produced in a given year to the base-year hydropower produced. As hydroelectricity represents 8 per cent of electricity production in Egypt (specifically, 7.7 per cent as of 2013: World Bank 2016), the hydropower loss or gain is born by 8 per cent of the electricity sector. The shock to TFP for electricity therefore cannot exceed 8 per cent of electricity production, even in the case of a large drought and rapid filling of the GERD. We neglect the impacts on thermo-electric cooling requirements, as they are a priority for water use and are coved by irrigation flows going to the Nile Delta.

With these impact parameters as inputs, the CGE model is solved a second time. During the second run, no changes in the allocation of crops by land are allowed, since farmers cannot change their decisions about what to grow once their crops are already planted. The second solution of the CGE model provides new equilibrium prices and quantities that represent the realized economic outcomes given the agricultural water and hydropower shocks. The model repeats this process for every year.

Results

The economy-wide results focus on the first three years of the GERD filling period when impacts to Egypt, and differences between fill policies, are most pronounced. Although differences between the scenarios persist for over a decade (see Figure 7.2), these grow considerably smaller after three years and, from an economic perspective, would be diminished further by discounting. We present the model output as physical quantities of reductions in HAD hydropower generation and Egypt irrigation water deliveries, and then compare these to the resulting reductions in Egypt's gross domestic product (GDP). These GDP impacts are then compared to the potential hydropower revenues of the GERD over the same three-year period. In each case, we report the '5 per cent risk level' and 'average effect' across the 100 stochastic runs. The 5 per cent risk level is the level of output (hydropower generation, water deliveries or GDP) that is exceeded 95 per cent of the time, or alternatively, it has a 5 per cent chance of having a worse outcome. These two cases are reported across the required GERD fill durations (unconstrained, three-year and ten-year), and annual release requirements (0 BCM, 15 BCM and 30 BCM).

Effect on Egyptian hydropower and irrigation water needs

In the first three years of filling, the reductions in HAD hydropower generation and irrigation water deliveries are highest under the unconstrained-0 BCM scenario, and lowest under the 10-year-30 BCM scenario (Table 7.1). Under the unconstrained-0 BCM fill scenario at a 5 per cent risk level, hydropower generation falls by 10.3 per cent and irrigation deliveries by 2.1 per cent on average each of the first three years. These translate to approximately 1 terawatt-hour (TWH) of electricity, and 1 BCM of irrigation water deliveries, each of which is a significant effect. If a ten-year fill policy is adopted, impacts to generation are 60 per cent lower and impacts to irrigation deliveries fall by 40 per cent. Results

Table 7.1 Average reductions in HAD hydropower generation and Egypt irrigation water deliveries over first three years of GERD filling, under three filling duration and release requirements

GERD fill duration	5% risk level			Average effects		
	0 BCM	15 BCM	30 BCM	0 BCM	15 BCM	30 BCM
HAD hydropower generation						
Unconstrained	10.3%	8.1%	5.4%	6.8%	5.9%	4.0%
Three-year	7.9%	7.3%	5.4%	6.0%	5.4%	4.0%
Ten-year	4.0%	3.9%	3.6%	3.0%	3.0%	2.9%
Irrigation water deliveries						
Unconstrained	2.1%	1.8%	1.4%	1.4%	1.1%	0.8%
Three-year	2.1%	1.8%	1.4%	1.3%	1.1%	0.8%
Ten-year	1.2%	1.2%	1.1%	0.6%	0.6%	0.5%

from Wheeler *et al.* (2016) are not directly comparable to these due to differences in streamflow time series, but they generally corroborate our results.

Effect on Egypt's GDP and wages

The CGE-W model produces changes in total GDP and agricultural GDP effect. In relative terms, the impacts on total GDP is over fifty times lower than the impact on hydropower generation (Table 7.2). In the most aggressive fill scenario at a 5 per cent risk level, Egypt's GDP falls by 0.13 per cent and agricultural GDP by 0.37 per cent over the first three years of filling. As with hydropower and irrigation deliveries, moving to longer fill durations and release requirements lowers the impacts to GDP.

Although the impacts on wages in Egypt over the first three years of GERD filling are modest, the largest negative effects are experienced by the poorest segments of the urban and rural populations (Figure 7.4). Wages in the poorest 20 per cent of urban and rural populations fall approximately 0.13 per cent on

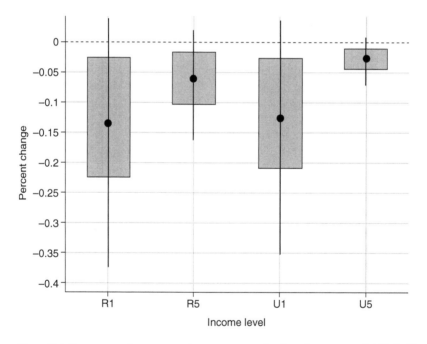

Figure 7.4 Percentage change in real income over the first three years of GERD filling, under the three-year filling duration and 15 BCM release scenario.

Source: own analysis of CGE-W outputs.

Notes
On the horizontal axis, R is rural and U is urban, and 1 is the bottom 20 per cent of incomes and 5 is the top 20 per cent of incomes. In the boxplot, the vertical line spans the 5th to 95th percentiles, the box from 25th to 75th percentiles, the open circle is the median (50th percentile), and the filled circle is the mean.

Table 7.2 Average reduction in Egyptian GDP and agricultural GDP over first three years of GERD filling, under three filling duration and release requirements

GERD fill duration		5% risk level			Average effects		
		0 BCM	15 BCM	30 BCM	0 BCM	15 BCM	30 BCM
Total GDP	Unconstrained	0.13%	0.12%	0.09%	0.10%	0.08%	0.05%
	Three-year	0.13%	0.12%	0.09%	0.09%	0.07%	0.05%
	Ten-year	0.08%	0.07%	0.07%	0.04%	0.04%	0.04%
Agricultural GDP	Unconstrained	0.37%	0.32%	0.26%	0.28%	0.22%	0.15%
	Three-year	0.35%	0.32%	0.26%	0.25%	0.21%	0.15%
	Ten-year	0.23%	0.22%	0.21%	0.11%	0.11%	0.10%

average and 0.35 per cent at the 5 per cent risk level. These changes are similar to reductions in agricultural GDP that drives a large fraction of employment in lower income categories. On the other hand, wages of the richest 20 per cent of the rural and urban population fall by only half and one-quarter as much, respectively. The magnitude and pattern of impacts are similar for the unconstrained and three-year fill durations and are much lower in the ten-year fill duration.

Trade-off: Egypt impacts and Ethiopia benefits

How do these economic impacts to Egypt compare to economic benefits to Ethiopia over the same three-year period? For context, once filled, GERD hydropower generation is modelled at approximately 12 TWH of electricity annually. Electricity prices in neighbouring Kenya, which could be a purchaser of GERD energy, were approximately $0.13 per kilowatt-hour (kWh; Regulus 2016). Applying a conservative price of $0.10 per kWh to this generation estimate gives annual revenues of $1.2 billion. This is a lower bound estimate of sustained GDP gains due to the multiplier effects that electrification would produce within the economy. Using this lower bound estimate of Ethiopian GDP gains, Table 7.3 compares Ethiopian gains to Egypt's losses over the first three years of filling, under the 15 BCM annual GERD release requirement scenario. Depending on the risk level and fill duration scenario, mean annual gains by Ethiopia range from 2.5 to 6 times Egyptian losses over this three-year period. In terms of fill policy, on average Ethiopia gains $400 million annually over the three years by rapidly filling the dam instead of imposing a ten-year fill requirement, whereas Egypt loses $110 million per year. After the third year, Ethiopian gains continue to rise while Egyptian losses fall. This large net benefit for the Nile system suggests room for cooperation between the two countries.

Discussion and further research

We find that GERD filling can reduce HAD hydropower generation and irrigation deliveries by up to 10 per cent and 2 per cent, respectively, but that these impacts are greatly dampened, in relative terms, when translated into GDP. There are two explanations for this. The first is that economies are elastic:

Table 7.3 Average annual Ethiopian hydropower gains and Egyptian GDP losses (millions of US$) over the first three years of GERD filling, assuming a 15 BCM release requirement

GERD fill duration	5% risk level		Average effects	
	Ethiopia gains	*Egypt losses*	*Ethiopia gains*	*Egypt losses*
Unconstrained	$894	$332	$1,043	$218
Three-year	$803	$332	$917	$207
Ten-year	$610	$209	$660	$108

substitution occurs and is reflected in the general equilibrium model. For example, in years where hydropower production falls, fossil fuel resources are substituted to avoid energy shortages or significant price increases. Second, the structure of Egypt's economy has changed significantly since the inauguration of the HAD in 1971. That year, hydropower generation made up 63 per cent of Egypt's energy generation, and 30 per cent of Egypt's GDP was from agricultural output. By 2013, structural change in Egypt's economy reduced hydropower to only 8 per cent of generation and agriculture to 11 per cent of GDP (Figure 7.5).

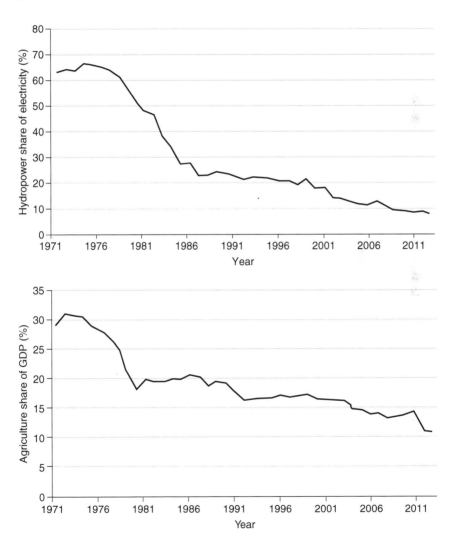

Figure 7.5 In Egypt: hydropower share of electricity (top) and agriculture share of GDP (bottom).

Source: World Bank 2016.

The impacts of GERD filling are a 'spin of the roulette wheel' that depends on the unknown magnitude and timing of near future Nile inflows. Even in light of this uncertainty, however, the worst-case impacts of unconstrained GERD filling on Egypt's economy are modest, suggesting that Egypt's economy is not significantly vulnerable to how GERD is filled. The much higher short-term gains to the Ethiopian economy mean that a more rapid fill policy would have higher Nile-wide economic benefits, opening the door for benefits sharing arrangements. Future research needs to expand this linked water-economic framework to include Ethiopia and Sudan in order to analyse the economy-wide implications of cooperative GERD-HAD management strategies. For the longer-term consequences of the GERD, it will also be critical to better understand how Sudan's water use and management will change once the GERD is constructed.

Acknowledgements

The authors wish to thank Arthur Gueneau and James Thurlow for technical contributions and project support, as well as Hans Lofgren and Moataz El-Said for their earlier work on the basic Egyptian CGE model. We gratefully acknowledge the financial support of the United Nations University World Institute for Development Economic Research (UNU-WIDER).

References

Allen, R., Droogers, P. and Hargreaves, G. (2002). Predicting Reference Crop Evapotranspiration with Arid Weather Data. Prepared for the International Commission on Irrigation and Drainage (ICID). Retrieved 1 May 2012, from www.kimberly.uidaho.edu/water/papers/index.html.

Block, P., Strzepek, K. and Rajagopalan, B. (2007). Integrated Management of the Blue Nile Basin in Ethiopia: Hydropower and Irrigation Modeling. IFPRI Discussion Paper 00700.

FAO (Food and Agricultural Organization of the United Nations) (1998). Crop Evapotranspiration – Guidelines for Computing Crop Water Requirements – FAO Irrigation and Drainage Paper 56. Retrieved 22 March 2011, from www.fao.org/docrep/x0490e/x0490e00.htm#Contents.

FAO. (2012). PriceStat. Retrieved 1 May 2012, from http://faostat.fao.org/site/570/default.aspx.

Howell, P., Allan, J. and Conway, D. (1995). The Nile: Sharing a Scarce Resource. An Historical and Technical Review of Water Management and of Economical and Legal Issues. *Global Environmental Change-Human and Policy Dimensions*, 5(2), 157.

Jeuland, M. (2010). Economic Implications of Climate Change for Infrastructure Planning in Transboundary Water Systems: An Example from the Blue Nile. *Water Resources Research*, 46, W11556. Doi: 10.1029/2010WR009428.

Keller, A. and Keller, J. (1995). *Effective Efficiency: A Water Use Efficiency Concept for Allocating Freshwater Resources*. Discussion Paper 22, Center for Economic Policy Studies. Winrock International.

Lofgren, H., Harris, R. and Robinson, S. (2001). *A Standard Computable General Equilibrium (CGE) Model in GAMS*. TMD Discussion Paper No. 75. Washington, DC: International Food Policy Research Institute.

Mobasher, A.M. (2010). *Adaptive Reservoir Operation Strategies Under Changing Boundary Conditions : The Case of the Aswan High Dam Reservoir*. PhD Dissertation, Technische Universität Darmstadt, Hessen.

Regulus (Regulus Limited) (2016). Electricity Cost in Kenya. Retrieved 15 November 2016, from https://stima.regulusweb.com/.

Robinson, S. and Gehlhar, C. (1996). Impacts of Macroeconomic and Trade Policies on a Market-Oriented Agriculture. In Lehman B. Fletcher (ed.), *Egypt's Agriculture in a Reform Era*. Ames, IO: Iowa State University Press.

Robinson, S. and Gueneau, A. (2013). *Pakistan Under Water Stress: An Economic Analysis of Some Proposed Water Policies Using CGE-W*. IFPRI Pakistan Strategy Support Program Working Paper.

Robinson, S., Strzepek, K., El-Said, M. and Lofgren, H. (eds) (2008). *Indirect Impact of Dams: Case Studies from India, Egypt, and Brazil* (pp. 227–273). Washington, DC, and New Delhi, India: World Bank and Academic Foundation.

Strzepek, K., Yohe, G., Tol, R. and Rosegrant, M. (2007). The Value of the High Aswan Dam to the Egyptian Economy. *Ecological Economics*, 66(1), 117–126.

Sutcliffe, J. and Parks, Y. (1999). *The Hydrology of the Nile*. IWMI. IAHS Special Publication no. 5.

Wheeler, K.G., Basheer, M., Mekonnen, Z.T., Eltoum, S.O., Mersha, A., Abdo, G.M., Zagona, E.A., Hall, J.W. and Dadson, S.J. (2016) Cooperative Filling Approaches for the Grand Ethiopian Renaissance Dam. *Water International*, 41(4), 611–634. Doi: 10.1080/02508060.2016.1177698.

Whittington, D., Wu, X. and Sadoff, C. (2005). Water Resources Management in the Nile Basin: The Economic Value of Cooperation. *Water Policy*, 7, 227–252.

World Bank (2016). *Indicators: Agriculture and Rural Development, and Energy and Mining*. Retrieved 15 November 2016, from http://data.worldbank.org/indicator.

Zhang, Y., Erkyihun, S.T. and Block, P. (2016). Filling the GERD: Evaluating Hydroclimatic Variability and Impoundment Strategies for Blue Nile Riparian Countries. *Water International*, 41(4), 593–610. Doi: 10.1080/02508060.2016.1178467.

8 Economic impact assessment of the Grand Ethiopian Renaissance Dam under different climate and hydrological conditions

Tewodros Negash Kahsay, Onno Kuik, Roy Brouwer and Pieter van der Zaag

Introduction

Ethiopia is building a large hydropower dam known as the Grand Ethiopian Renaissance Dam (GERD) on the Blue Nile River close to the border with Sudan. The dam constitutes the centrepiece of the country's five-year Growth and Transformation Plan (GTP) (2010/11–2014/15) that aims to boost the country's hydropower generating capacity from 2,000 MW in 2009/10 to 10,000 MW in 2014/15 (MoFED 2010). The GERD is estimated to cost €3.34 billion and will have a height of 145 metres and a total storage volume of 74 km³. The dam, which is planned to be used for power generation only, has a design capacity of 6,000 MW and is reported to be able to produce 15.1 TWh/ year upon completion (MDI 2012). This would mean a massive additional energy source in the country and is expected to create enough supply to meet domestic as well as export demand for electricity.

The Blue Nile River constitutes the most important source of water supply in downstream countries Sudan and Egypt. The project has therefore been a source of concern for these downstream countries. Ethiopia argues that the GERD will offer several benefits to these countries, including hydropower supply at a comparably cheaper price, flood control, water savings through reduced evaporation losses from downstream reservoirs and trapping silt. In order to create trust and consensus on the dam, the Eastern Nile countries agreed on the establishment of an International Panel of Experts (IPoE), tasked with assessing the impact of the dam on downstream countries. The IPoE's report indicates, among other things, that the dam could potentially offer significant benefits to all the three Eastern Nile countries (MoFA 2013). This corresponds with findings of recent studies on the dam (Kahsay et al. 2015; Arjoon et al. 2014) as well as earlier studies that assessed the downstream impact of a cascade of hydropower projects with a combined storage capacity comparable to that of the GERD on the Ethiopian part of the Blue Nile Basin (e.g. Guariso and Whittington 1987; Whittington et al. 2005).

The economic impact of major hydraulic infrastructure projects like dams has been analysed using both partial and general equilibrium models. Partial equilibrium analysis incorporates detailed information on the hydraulic infrastructure and the hydrologic processes. Such analysis contains relevant details concerning land and water use and crop production at scheme level and hydropower generation at power plant level, but lacks relevant economic feedbacks due to its partial equilibrium nature. Several partial equilibrium models have been developed to analyse the economic effects of infrastructure development on the Ethiopian part of the Blue Nile River (e.g. Guariso and Whittington 1987; Whittington *et al.* 2005; Blackmore and Whittington 2008; Goor *et al.* 2010; Block and Strzepek 2010; Arjoon *et al.* 2014). Their findings reveal that Ethiopia's development of the Blue Nile waters to maximise hydropower production would not have a significant effect on water supply to downstream countries.

Computable General Equilibrium (CGE) models consider the entire economy as an interdependent system, therefore providing an economy-wide perspective. Unlike partial equilibrium models, CGE models account for various inter-linkages between economic sectors to analyse economy-wide effects that could occur as a consequence of policy interventions such as big dams. Nevertheless, there are only a few studies in the literature that examine the economic effects of dams using CGE models (e.g. Strzepek *et al.* 2008; Wittwer 2009; Ferrari *et al.* 2013). These studies disregard the transient phase of impounding reservoirs and are confined in scope to the analysis of national impacts of dam projects. A recent CGE study by Kahsay *et al.* (2015) evaluates the transboundary economic impacts of the GERD, taking account of the integrative nature of the river basin. The authors used the GTAP-W model to evaluate the impact of the dam under three different climate and hydrological scenarios, taking into account both the transient impounding phase and the long-term operation phase of the dam. The results demonstrate the significance of the GERD in generating basin-wide economic benefits and improving welfare in the Eastern Nile Basin.

The study presented here employs the same CGE methodology to assess the potential transboundary economic impacts of the GERD on the Eastern Nile Basin countries. The economic effects of the dam in the basin are analysed under three different climatic and hydrological conditions, taking into account both the transient impounding phase and the long-term operation phase of the dam using the revised version of the GTAP-W model (Calzadilla *et al.* 2010). The study is based on the assumptions and the scenarios implemented in the study reported by Kahsay *et al.* (2015), but the analysis presented here is based on a new data set, using the latest GTAP 9 database (Aguiar *et al.* 2016).

The remainder of the chapter is organised as follows. The next section briefly reviews existing models applied to the economic analysis of dams. Section three presents the theoretical modelling framework and the data aggregation procedure. Section four introduces the simulation scenarios. Section five discusses the results and section six concludes.

Economic models of dams

In the literature, the economic impacts of dams have been assessed using partial and general equilibrium modelling approaches. A partial equilibrium model generally relies on a network representation of the water resources system to physically connect water demands with supplies. The model reproduces the physical behaviour of the river system using a detailed representation of surface water resources and water demand nodes, including off-stream (irrigation fields) and in-stream (hydropower) demand, which are spatially connected to the river basin network. The model, which encompasses the entire river network, hydraulic infrastructure, water and land demand in irrigated agriculture as well as detailed hydrologic data, is used to derive the optimal allocation of land and water across crops in agricultural production and optimal hydropower generation using hydropower dams. The strength of a partial equilibrium model lies in its ability to incorporate detailed information about the hydraulic infrastructure and associated hydrologic processes. However, due to the partial equilibrium nature of the model, the results represent certain sectors of an economy, typically agriculture and hydropower, and are confined to the hydrological area, which is not connected to the economy in the non-hydrological area. The model thus depicts only part of the overall economy and assumes that there exist no inter-sectoral linkages and relevant economic feedbacks. Moreover, partial equilibrium models are incomplete in the sense that they fail to capture the potentially important relationships between prices and quantities. Prices are treated as exogenous while, in practice, changes in outputs are expected to affect product prices and changes in input demands could well be reflected in input prices.

Several studies analysed the economic effects of infrastructure development in the Blue Nile River in Ethiopia using partial equilibrium models (e.g. Jeuland 2010a, 2010b; Whittington *et al.* 2005; Block and Strzepek 2010; Goor *et al.* 2010). These studies are essentially deterministic and assume perfect foresight. Only Goor *et al.* (2010) adopt a stochastic programming approach to assess the economic benefits and costs associated with new water storage infrastructure in the upper Blue Nile in Ethiopia. The findings of this study reveal that the construction of four mega dams (Karadobi, Beko-Abo, Mandaya and Border) in the Blue Nile Basin in Ethiopia would have tremendous positive impacts on hydropower generation and irrigation in Ethiopia and Sudan. Moreover, evaporation losses from the reservoir of the High Aswan Dam would be reduced substantially if operation of the reservoirs were coordinated. Similarly, a recent partial equilibrium analysis of the economic impact of the GERD presented by Arjoon *et al.* (2014), applying a stochastic dual dynamic programming (SDDP) approach to assess the impact of the operation of the dam on the Eastern Nile economies, shows that water storage in the GERD would benefit downstream countries through improved irrigation and hydropower development and reduced hydrologic risks, particularly during dry years.

Unlike partial equilibrium models, which analyse the different sectors separately under *ceteris paribus* assumptions, the general equilibrium approach accounts for various inter-linkages between economic sectors to analyse economy-wide effects that could occur as a consequence of a policy change. CGE models are best suited to analyse the direct as well as indirect impacts of large-scale policy interventions such as big dams on interconnected economic systems (Robinson *et al.* 2008). Due to their general equilibrium feature, CGE models account for various inter-linkages between economic sectors to analyse economy-wide effects that could result from a policy change. Moreover, CGE models determine relative product and factor prices endogenously so that product and factor markets attain their equilibrium through the adjustment of prices. Compared to partial equilibrium models, CGE models are highly aggregated and lack the relevant hydrological details needed to assess the optimal spatial allocation of water and land resources across crops at agricultural plot level and hydropower generation at plant level.

Several studies have used CGE models to examine a wide range of water related issues (e.g. Seung *et al.* 1998; Seung *et al.* 2000; Diao and Roe 2003; Gomez *et al.* 2004; Diao *et al.* 2005; Feng *et al.* 2007; Brouwer *et al.* 2008; Van Heerden *et al.* 2008). However, CGE analyses of water infrastructure are not that common and their scope is typically confined to national impacts of dam projects. Examples include Egypt's High Aswan Dam (HAD) (Strzepek *et al.* 2008), Australia's Traveston dam (Wittwer 2009) and Ethiopia's GERD (Ferrari *et al.* 2013). Strzepek *et al.* (2008) used a CGE model of the Egyptian economy to estimate the economic benefits of the HAD from an economy-wide perspective. They conducted a comparative-static simulation of the Egyptian economy with and without the HAD. Their results show that the HAD benefits Egypt by considerably increasing its GDP. Employing a dynamic version of the GOBE_EN CGE model, Ferrari *et al.* (2013) conducted a preliminary assessment of the possible economic effects of the GERD on the Ethiopian economy. Their results indicate that the investment in the GERD would slow down development in Ethiopia and exports of hydroelectricity would need to expand rapidly after dam completion for the investment to be profitable. They furthermore argue that an increase in the Ethiopian export of electricity could lead to 'Dutch disease', that is, appreciation of real exchange rate, leading to a decline in the country's current exporting sectors by making them less competitive on the export market.

Previous CGE studies typically disregard the transient phase of impounding reservoirs. Moreover, they focus on national impacts disregarding the integrative transboundary nature of the river basin on which the dam is constructed (e.g. Strzepek *et al.* 2008; Ferrari *et al.* 2013). A recent study by Kahsay *et al.* (2015) represents one of the first efforts to use a global CGE model to analyse the transboundary economic impacts of the GERD taking into account both the impounding and operation stages across the whole Nile River Basin.

Modelling framework and data

Modelling framework

The CGE model applied in this study is based on the Global Trade Analysis Project (GTAP) model (Hertel 1997), developed at Purdue University, USA. GTAP provides a global modelling framework and a common global database, providing the opportunity to conduct comparable model implementation and policy simulations. GTAP is a static-comparative, multi-region, multi-sector CGE model of the world economy that examines all major aspects of an economy via its general equilibrium feature. The GTAP model comprises accounting relationships, behavioural equations and global sectors required to complete the model. The accounting relationships of the model ensure the balance of receipts and expenditures for every agent identified in the economy, whereas the behavioural equations specify the behaviour of optimising agents in the economy on the supply and demand side based on microeconomic theory (Brockmeier 2001). The production system is set up as a series of nested constant elasticities of substitution (CES) functions combined through substitution elasticities.

The analysis presented here uses the version of the GTAP-W model (Calzadilla *et al.* 2010). The value of water in the model is assumed to be embedded in the value of land (Calzadilla *et al.* 2011). The GTAP-W model (Calzadilla *et al.* 2010) incorporates a water module in the GTAP model based on the GTAP-E version developed by Burniaux and Truong (2002) for the analysis of energy markets and environmental policies. To account for water, the agricultural land endowment in the standard GTAP-E database is disaggregated into rain-fed land, irrigable land, and irrigation water based on baseline data on irrigated and rain-fed agriculture. The relative share of rain-fed and irrigated production in total production is used to split the land rent in the GTAP-E database into a value for rain-fed land and a value for irrigated land for each crop in each region. In a next step, the ratio of irrigated yield to rain-fed yield is used to split the value of irrigated land into the value of irrigable land and the value of irrigation water. This results in a global CGE model that distinguishes between rain-fed and irrigated agriculture and incorporates water as a factor of production directly substitutable in the production process of irrigated agriculture (see Figure 8.1).

In the GTAP-W model, as in the standard GTAP model, primary factors of production are assumed to substitute for one another according to a constant substitution elasticity parameter. The substitution elasticity parameter, which defines the relationship between changes in the ratio of factor inputs used in the production of a given level of output and the inverse ratio of their marginal products, that is, their inverse price ratio in equilibrium describes the flexibility of a production technology to allow for changes in the quantity ratios of factors used in the production of a given level of output as relative factor prices change.

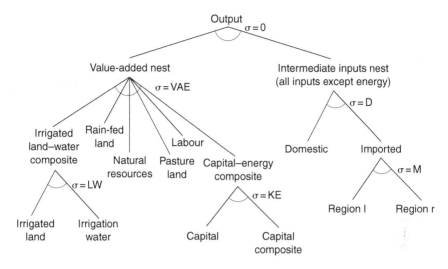

Figure 8.1 Truncated nested production structure in the GTAP-W model.
Source: Calzadilla *et al.* (2010).

Data

The GTAP 9 Data Base is used for the study. The GTAP 9 Data Base includes 140 regions and sixty-eight commodities and features three reference years (2004, 2007 and 2011) (Aguiar 2016). More specifically, the GTAP-Power 9 Data Base is used, an electricity-detailed extension of the GTAP 9 Data Base, which disaggregates the electricity sector in the GTAP 9 Data Base into twelve sub-sectors (Peters 2016). The latest reference year of the database (2011) is used as a baseline for the study, and the 140 regions are aggregated into eight regions: Egypt, Ethiopia, Sudan (pre-2011), the Equatorial Lakes (EQL) region, Rest of East Africa, Rest of North Africa, Rest of Sub-Sahara Africa and Rest of the World (ROW). The four EQL countries covered in the GTAP 9 Data Base are Rwanda, Kenya, Tanzania and Uganda. In the GTAP 9 database, Sudan is aggregated in the Rest of Eastern Africa region. The SplitReg Program (Horridge 2011a) is employed to split off Sudan from the composite region based on the share of Sudan in the region's total value of endowments. The split database is verified using the GTAPAdjust Program (Horridge 2011b). Since the study focuses on water resources management of the Nile River Basin, the regional aggregation highlights the importance of the Eastern Nile region, where most of the Nile water resources are generated and used.

For the purpose of this study, the sixty-eight sectors in the GTAP-Power 9 Data Base are aggregated into twenty sectors, of which eight are agricultural and twelve non-agricultural sectors. The electricity sub-sectors in the Data Base are aggregated into three (hydropower, fossil and other sources). Following Calzadilla *et al.* (2011), the agricultural land endowment in the standard GTAP database is disaggregated into rain-fed land, irrigable land and irrigation water, based on data generated by the IMPACT model (Nelson *et al.* 2010). The relative share of rain-fed and irrigated production in total production is used to split the land rent in the original GTAP database into a value for rain-fed land and a value for irrigated land for each crop in each region. Due to the lack of data, the values for the elasticity of substitution between irrigated land and irrigation water used in this study are adapted from Calzadilla *et al.* (2011), and vary between 0.05 in Ethiopia and Sudan and 0.08 in Egypt. Sensitivity analysis of model results with respect to these elasticity parameters is conducted in the sensitivity analysis section below.

Scenarios

The economic impacts of the GERD are assessed under a number of alternative scenarios. For the impounding phase, the scenarios distinguish between three hydrological conditions (average, wet and dry) and two water withdrawal options in Sudan (low and high). The combinations of these conditions and options provide six scenarios (SS1 to SS6). A seventh scenario (SS7) presents our main assumptions governing the operational phase of the dam. The key assumptions underlying the seven scenarios are presented in Table 8.1.

The GERD impounding phase

For the impounding phase of the GERD, the scenarios distinguish between three possible hydrological conditions. To represent average hydrological conditions, we take the average of the hydrological conditions observed over the period 1970–1978. Representative of wet conditions is the average over the period 1993–2001, when the inflow of water at both the GERD and the HAD was 14 per cent larger than under average conditions (Tractebel Engineering and GDF Suez 2011). Representative of dry conditions are the hydrological conditions that were observed over the period 1979–1988, with a 20 per cent lower inflow at both the GERD and the HAD.

Two options are considered for water withdrawal in Sudan. The first option implies a water withdrawal rate of $15\,km^3$ per year that is slightly above the existing water withdrawal rate of $13.8\,km^3$ (Blackmore and Whittington 2008). The second option is a water withdrawal rate of $18.5\,km^3$ per year, which is the maximum rate allowed for Sudan according to the 1959 agreement between Egypt and Sudan. A rate of $18.5\,km^3$ per year may be considered high in the short-term impounding phase considering the current storage capacity and existing irrigation infrastructure, but the option enables a sensitivity analysis that captures the maximum impact of the GERD impounding on the economy of Egypt.

The potential impacts of the GERD impounding on the HAD in terms of hydropower generation and irrigation water supply in Egypt for the three hydrological regimes and Sudanese water withdrawal options are presented in Table 8.2.

Table 8.1 Main assumptions underlying the future change scenarios related to the GERD (percentage change compared to the 2011 baseline conditions)

| | GERD impounding | | | | | | GERD operation |
	SS1	SS2	SS3	SS4	SS5	SS6	SS7
Hydrological conditions	Average	Wet	Dry	Average	Wet	Dry	Average
Ethiopia							
Hydroelectric production	+122.0	+149.0	+104.0	+122.0	+149.0	+104.0	+237.0
Unskilled labor	+5.5	+5.5	+5.5	+5.5	+5.5	+5.5	0
Domestic saving	+10.0	+10.0	+10.0	+10.0	+10.0	+10.0	0
Capital stock	0	0	0	0	0	0	+10
Sudan (pre-2011)							
Water withdrawal (km³/yr)	15	15	15	18.5	18.5	18.5	18.5
Hydroelectric production	+7.0	+7.0	+7.0	+7.0	+7.0	+7.0	+35.4
Capital stock	+1.3	+1.3	+1.3	+1.3	+1.3	+1.3	+3.0
Irrigation water supply	0	0	0	+23.0	+23.0	+23.0	+23.3
Irrigated land	0	0	0	+18.0	+18.0	+18.0	+23.3
Water use efficiency	0	0	0	0	0	0	+5.0
Egypt							
Hydroelectric production	-8.0	-14.0	-18.0	-10.0	-6.0	-18.0	-8.0
Irrigation water supply	0	0	-6.0	0	0	-11.0	+3.6
Water use efficiency	0	0	0	0	0	0	+5.0

Table 8.2 GERD average power generation during the impounding stage, water use in Sudan and influence on the HAD

Climatic and hydrological scenarios	Mean GERD energy generation (TWh/yr)	Water use in Sudan (Km³/yr)	HAD energy generation		
			Without GERD (TWh/yr)	With GERD (TWh/yr)	Difference (%)
Average years	7.8	18.5	8.1	7.3	−10
		15			−8
Dry years	6.7	18.5	6.4	5.2	−18
		15			−18
Wet years	9.6	18.5	9.1	8.5	−6
		15			−14

Under average or wet hydraulic conditions, full impounding can be achieved in six years (2014–2019). This is not expected to have an impact on the irrigation water supply in Egypt, but energy generation in Egypt is adversely impacted due to a reduced reservoir level in the HAD. Under dry hydraulic conditions, the GERD impounding will take one more year and is expected to be achieved in seven years (2014–2020). The GERD impounding during a sequence of dry years is expected to have adverse effects on water supply and energy production in Egypt.

Power production at the HAD in Egypt decreases between 6 and 18 per cent, depending on the hydrological conditions and the water withdrawal rates in Sudan. Presently, water spills at the HAD during wet years. In these wet years, the excess water is used to generate additional hydropower (MDI 2012). With the impounding of the GERD upstream, this excess water is used to fill the GERD reservoir. As a consequence, the water level in the HAD will rarely exceed its normal level during the impounding phase of the GERD and hence outflows will be lower at the HAD and less hydropower can be produced (Tractebel Engineering and GDF Suez 2011). This effect is so strong that, in the case of lower water withdrawals in Sudan (SS1 to SS3 in Table 8.1), the impact of the impounding of the GERD on the HAD hydropower generation is larger in wet than in average years. Obviously, the greatest negative impact on hydropower generation occurs in dry years when there is also a decrease of 6 per cent in irrigation water supply in Egypt. At higher Sudanese water withdrawal rates (SS4 to SS6 in Table 8.1), power generation at the HAD decreases by 6 to 18 per cent and in dry years, irrigation water supply in Egypt decreases by 11 per cent.

The aggregate impacts of the GERD on Egypt that we estimate in this study are relatively high compared to estimates from previous studies. Asegdew and Semu (2014) estimate that the GERD impounding reduces the HAD power generation by 12.7 per cent on average, and that it has no effect on irrigation water supply. In a similar vein, Asegdew et al. (2014) estimate a fall in power generation of 3 to 17 per cent between 2015 and 2019. In contrast, a substantial increase in irrigation water supply (18 per cent) in Egypt is expected because of reduced evaporation in the HAD (Asegdew and Semu 2014; Asegdew et al. 2014). Our estimates of the adverse effects on power production and irrigation water supply in Egypt are therefore somewhat higher and are therefore not likely to *underestimate* the impact of the GERD on the HAD.

It has been estimated that the impounding of the GERD will reduce downstream sediment flows and sediment loads in the reservoirs of the Roseires, Sennar and Merowe hydropower plants in Sudan. This may increase power generation by these plants by 6.8 per cent from 8.8 TWh/year to 9.4 TWh/year (Asegdew et al. 2014). Avoided damage due to sedimentation (e.g. floods, dredging costs, reduced lifespan of dams) are assumed to increase Sudan's effective capital stock by 1.3 per cent per annum in the impounding phase (SS1 to SS6 in Table 8.1).

The GERD itself is expected to generate 6.7 to 9.6 TWh per year in the impounding phase depending on the hydrological conditions (Table 8.1). This corresponds to an increase of hydropower generation in Ethiopia of 104 per cent (dry years) to 149 per cent (wet years) (Table 8.1). Construction of the GERD is expected to increase the employment of unskilled labour by about 5.5 per cent during the four to five years it takes to construct the dam. Because of Ethiopia's high rates of rural underemployment (FAO 2013) and urban unemployment (urban unemployment was estimated to be 19.4 per cent in 2010 [ILO 2013]), we assume that the additional labour demand can be met from this under-employed and unemployed labour reserve.

The construction of the dam requires investments, which are assumed to correspond to a 10 per cent annual increase in domestic savings. We assume that the investment is financed by the issuing of bonds by the government of Ethiopia. We also assume that this investment does not reduce investments in other capital projects in Ethiopia that are largely financed by multilateral and bilateral foreign loans and foreign direct investments.

The GERD operation phase

The GERD reservoir is expected to be filled in 2019 or 2020 depending on whether impounding occurs during an average, wet or dry sequence of years. When the GERD becomes operational after the reservoir has been filled, it will no longer reduce the water flow downstream. While evaporation at the GERD reservoir will increase ($1.7 \, km^3$/year; [MDI 2012]), evaporation at the HAD reservoir is expected to decrease. Blackmore and Whittington (2008) estimate that full development of the Ethiopian part of the Blue Nile hydropower poten-tial, that is, the construction of a series of dams with a combined reservoir storage capacity equivalent to that of the GERD, would reduce the HAD evapo-ration losses by $1.9 \, km^3$/year.

The GERD is expected to generate 15.1 TWh per year. This corresponds to an increase of Ethiopia's hydropower production by 237 per cent (SS7 in Table 8.1). Because of reduced sediments flows, hydropower generation in Sudan is expected to increase by 35.5 per cent from 8.5 TWh/year to 11.6 TWh/year (Table 8.3). The GERD is also expected to mitigate flood damage to agriculture and infrastructure in Sudan from the Ethiopia–Sudan Border to Khartoum by up to US$200 million per annum (MDI 2012). Flood damage will also be mitigated in cities like Dongola, located far north of Khartoum (ENTRO 2006). Reduced sedimentation flows also mitigate damage to Sudan's irrigation canals and equip-ment, reduce reservoir and canal dredging costs, and increase the lifespan of the Sudanese dam reservoirs. Overall, flood damage reduction, reduced sedimenta-tion and the increased lifespan of dams are assumed to modestly augment Sudan's capital stock by 3 per cent in the GERD's operational phase (SS7 in Table 8.1).

The GERD will regulate the water flow in the Blue Nile River and the main Nile River, and will therefore generate a more constant flow over the year for irrigated agriculture and hydropower production in Sudan and Egypt. This would ensure a more continuous irrigation water supply and improved water use

Table 8.3 The expected impact of the GERD operation on downstream power generation in Sudan and the High Aswan Dam in Egypt (TWh/year)

	Without GERD	*With GERD*	*Difference (%)*
Sudan			
Roseires	2.2	3.4	+58
Sennar	0.1	0.1	+6
Merowe	6.2	8.0	+28
Total Sudan	8.5	11.5	+35
Egypt			
High Aswan Dam	8.0	7.4	−8

Source: Asegdew *et al.* (2014).

efficiency than currently is the case without the GERD. This will help realise Sudan's and Egypt's plans to expand their irrigation areas by 1.2 and 1.1 million hectares, respectively (ENTRO 2009). Reduced evaporation losses in the HAD and higher water withdrawal in Sudan (up to the maximum allowable rate of 18.5 km³ per year) and a slight increase in irrigation efficiency would allow Egypt and Sudan to realize a substantial share of their irrigation expansion plans. We estimate that this expansion increases total irrigation water supply by 3.6 per cent in Egypt and 23.3 per cent in Sudan. An increase in water use efficiency in irrigated agriculture is assumed in Sudan and Egypt only, not in Ethiopia, because irrigated agriculture is still very limited in Ethiopia (Tesfaye *et al.* 2016). On the downside, in the operational phase, the GERD would reduce power generation of the HAD by 8 per cent due to the loss of headwaters caused by the construction of the GERD (SS7 in Table 8.1).

Simulation results

The simulation results reveal that the GERD impounding has no effect on irrigation water supply in Ethiopia. Similarly, water endowments and hence water use across crops remain stable in Sudan during the GERD impounding if it maintains its current level of water withdrawal (SS1–SS3). If Sudan increases its water use during the impounding stage (SS4–SS6), its water endowment increases, resulting in a 16 to 28 per cent increase in water allocation across crops. The GERD impounding would have no effect on water use in Egypt, if it occurs during average and wet years. If it occurs during a sequence of dry years (SS3 and SS6), the GERD impounding entails a 2 to 15 per cent decline in water use across crops in Egypt depending on the level of water withdrawal in Sudan. The GERD operation induces significant positive effects on water use in downstream countries (Figure 8.2). When the GERD starts operating, water use increases substantially across agricultural sectors in Sudan (15–29 per cent), and to a lesser extent in Egypt (3–6 per cent). Although the GERD operation tends to increase water use in some sectors and decrease it in others in Ethiopia,

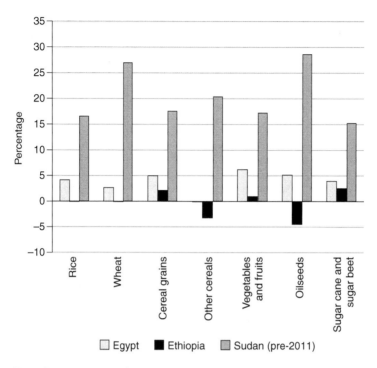

Figure 8.2 Percentage change in irrigation water allocation across agricultural sectors compared to the baseline scenario due to the GERD operation.

its net effect on irrigation water use in the country is negligible, since the dam is planned for hydropower generation only.

Changes in irrigation water use due to the GERD impounding and operation induce similar changes in agricultural production (Table 8.5). Following the pattern of changes in irrigation water supply, agricultural production remains stable in Egypt when impounding occurs during average and wet years, and declines in Egypt when occurring during a sequence of dry years (SS3 and SS6). The adverse effect of the GERD impounding on Egypt's agriculture is more pronounced, if Sudan increases its water withdrawal during dry years (SS6). In that case, Egypt is expected to experience a 0.1 to 5 per cent decline in agricultural production. The effect of the GERD impounding on the HAD power generation and subsequently on Egypt's manufacturing sector is found to be negligible (-0.01 per cent), since the country heavily relies on non-hydropower sources of energy. Although there is hardly any effect on irrigation water supply in Ethiopia, the GERD impounding is expected to improve agricultural output in Ethiopia (0.3–3.4 per cent) due to its positive effect on energy and labour supply. Similarly, the positive effect of the GERD impounding on power generation and the capital stock yields a slight improvement in agricultural production in

Sudan (0.1–0.4 per cent) under the country's current level of water withdrawal (SS1–SS3). If it increases its water withdrawal during the GERD impounding stage (SS4–SS6), Sudan enjoys considerable improvements in its agricultural output (1–11 per cent). The GERD impounding has limited and mixed effects on manufacturing output in Ethiopia and Sudan.

Agricultural production increases in all the Eastern Nile countries due to the GERD operation (SS7 in Table 8.5). Evaporation losses from the HAD reservoir are reduced due to the GERD operation. When the GERD starts operating, Egypt is expected to gain water from this reduced evaporation loss and benefit from improved irrigation efficiency due to a year-round regulated flow of the Nile, which enhances irrigated agriculture. As a result, production increases by 0.1 to 2.4 per cent in Egypt's agricultural sectors once the GERD starts operation. However, output decreases slightly (0.1 per cent) in the manufacturing sector due to a lower power generation from the HAD influenced by the GERD operation.

The GERD operation improves agricultural production in Sudan due to increased water supply and improved irrigation efficiency and the consequent expansion of irrigated agriculture. Sudan also slightly gains an increased manufacturing production (0.2 per cent) due to an increased capital stock and power generation in its power plants as sediment loads decrease with the GERD operating upstream. Due to the immense increase in power generation and gain in capital stock, the GERD operation enhances production in most of Ethiopia's agricultural sectors by 0.5 to 3.7 per cent and boosts the manufacturing production by 7 per cent.

The GERD generates substantial basin-wide improvements in real GDP. During the GERD impounding stage (SS1–SS6), Ethiopia and Sudan are expected to gain an increase in real GDP of US$911 million to US$963 million, and US$164 million to US$211 million in 2011 price levels, respectively, while Egypt faces a loss in real GDP of US$12 million to US$87 million. On average, Ethiopia and Sudan earn US$935 million and US$187 million, respectively, of the total basin-wide gain in real GDP of US$1.08 billion due to the GERD impounding, while Egypt loses US$38 million. Depending on the amount of water withdrawal in Sudan, the GERD impounding costs Egypt even more (US$63 million to US$87 million of its GDP), if it occurs in a dry sequence of years (SS3 and SS6). As the GERD enters its operation phase, the basin-wide gain in real GDP due to the GERD operation rises to about US$2 billion, of which Ethiopia, Sudan and Egypt earn US$1,474 million, US$448 million and US$75 million respectively. Thus, the GERD enhances real GDP in all the Eastern Nile countries when it starts operating. However, the distribution of the benefits is skewed with Ethiopia, Sudan and Egypt earning 74 per cent, 22 per cent and 4 per cent of the total basin-wide gain in real GDP, respectively.

The simulation results reveal that the GERD stimulates economic growth in the Eastern Nile Basin (Table 8.5). The average changes in real GDP in the Eastern Nile economies due to the GERD impounding (SS1–SS6) translate into economic growth rates of 3 per cent, 0.44 per cent and –0.02 per cent for

Table 8.4 The effect of the GERD impounding and operation stages on agricultural and manufacturing production (percentage change relative to the baseline scenario)

	Rice	Wheat	Cereal grains	Other cereals	Vegetables and fruits	Oilseeds	Sugar cane and beet	Livestock and meat	Manufactured products
SS1									
Egypt	0	0	0	0	0	0	0	0	-0.01
Ethiopia	1.33	2.65	3.27	1.29	2.96	0.34	3.14	3.45	-0.10
Sudan (pre-2011)	0.44	0.10	0.35	0.35	0.29	0.09	0.40	0.37	0.40
SS2									
Egypt	0	0.01	0	0	0	0	0	0	-0.01
Ethiopia	0.82	2.44	3.4	0.78	3.02	-0.32	3.23	3.46	-0.49
Sudan (pre-2011)	0.43	0.08	0.35	0.36	0.34	0.09	0.40	0.36	0.40
SS3									
Egypt	-0.03	-2.46	-0.02	-0.02	0.06	-0.13	-0.03	-0.05	0.04
Ethiopia	1.69	2.81	3.17	1.65	2.91	0.79	3.07	3.44	0.16
Sudan (pre-2011)	0.44	0.12	0.35	0.35	0.26	0.1	0.4	0.37	0.4
SS4									
Egypt	0	0	0	0	-0.01	-0.04	0	0	-0.01
Ethiopia	1.53	2.76	3.25	1.48	2.72	0.47	3.13	3.47	-0.02
Sudan (pre-2011)	0.83	10.77	1.14	4.20	2.84	9.68	0.74	0.55	-0.18
SS5									
Egypt	0	0	0	0	-0.01	-0.04	0	0	0
Ethiopia	1.01	2.55	3.39	0.96	2.78	-0.18	3.23	3.48	-0.41
Sudan (pre-2011)	0.83	10.75	1.14	4.21	2.89	9.68	0.74	0.55	-0.18
SS6									
Egypt	-0.08	-5.02	-0.06	-0.06	0.11	-0.37	-0.07	-0.10	0.10
Ethiopia	1.90	2.92	3.15	1.84	2.67	0.93	3.06	3.46	0.23
Sudan (pre-2011)	0.84	10.81	1.14	4.19	2.80	9.70	0.75	0.55	-0.18
SS7									
Egypt	0.51	2.37	0.87	0.50	0.05	1.70	0.23	0.45	-0.12
Ethiopia	0.57	0.45	3.06	-3.32	1.81	-5.04	3.74	3.42	6.96
Sudan (pre-2011)	1.43	13.25	1.74	5.64	4.21	11.75	1.32	1.04	0.19

Table 8.5 The effect of the GERD impounding and operation on real GDP (percentage change compared to the baseline scenario)

	SS1	SS2	SS3	SS4	SS5	SS6	SS7
Egypt	−0.01	−0.01	−0.03	−0.01	−0.01	−0.04	0.03
Ethiopia	2.99	3.09	2.93	2.99	3.08	2.92	4.73
Sudan (pre-2011)	0.38	0.38	0.38	0.49	0.49	0.49	1.05

Ethiopia, Sudan and Egypt, respectively. The GERD impacts positively on the Ethiopian and Sudanese economy throughout the impounding phase. The growth the GERD induces during the transient impounding stage in the Ethiopian economy is mainly due to changes in hydropower supply and employment of unskilled labour. Enhanced power generation in power plants due to reduced sediment loads and enhanced capital stocks due to reduced flood damage constitute the major factors that stimulate economic growth in Sudan during the GERD impounding stage. Increased water withdrawal during the GERD impounding (SS4–SS6) further improves economic growth in Sudan. The impact of the impounding stage on Egypt's economy is negligible. On average, Egypt's economy contracts by a mere 0.02 per cent during the GERD impounding phase. If it occurs during average and wet years, the GERD impounding affects the HAD hydropower generation only and results in an estimated 0.01 per cent decline in Egypt's economy. If the impounding occurs during dry years, Egypt faces a fall in power generation as well as a decline in irrigation water supply and hence the adverse effect on Egypt's economy increases slightly to 0.03 per cent to 0.04 per cent depending on the level of Sudan's water withdrawal. The simulation results reveal that 67 per cent (of the 0.3–0.4 per cent) of the decline in Egypt's economy is due to a fall in power supply as a result of the expected reduction in the HAD power generation. The remaining effect is attributable to reduced irrigation water supply. Thus, Egypt's economy is constrained more by the supply of energy than water.

The economies of all the Eastern Nile countries expand at a higher rate when the GERD goes operational. The Ethiopian economy expands at a rate of 4.7 per cent during the GERD operation. The long-term economic growth expected in Ethiopia due to the GERD operation emanates mainly from the combined change in the capital stock and hydropower supply. With higher benefits in terms of power generation and reduced flood risk damage, the GERD generates an economic growth rate of 1.1 per cent in Sudan during its operation phase. The GERD operation offers Egypt benefits too. With the GERD operating upstream, evaporation losses from the HAD reservoir decline, resulting in increased water supply and hence expanded irrigated agriculture. Moreover, a more regulated flow of water throughout the year due to the GERD operation provides an opportunity for improving water use efficiency in Egypt's irrigated agriculture. The GERD operation would therefore improve Egypt's economy by 0.03 per cent.

Since it tends to improve the real return to unskilled labour in all the Eastern Nile countries, the role of the GERD in alleviating poverty in the basin is significant. The real return to unskilled labour measures the change in return to unskilled labour relative to the price index of consumption expenditures and hence reflects trends in poverty reduction. The real return to unskilled labour is estimated to improve by 3.7 per cent, 1.9 per cent and 0.3 per cent in Ethiopia, Sudan and Egypt, respectively, due to the GERD operation. The simulation results thus reveal the significance of the GERD in reducing poverty, mainly in Ethiopia and Sudan and to some extent in Egypt. The expected contribution of the GERD in poverty alleviation in the Eastern Nile Basin is also reflected in the results for household income and consumption expenditures (Figure 8.3). Ethiopia is expected to see substantial improvements in household income (3.7–5.5 per cent) and hence consumption expenditures (3.7–5.2 per cent) throughout the GERD impounding and operation phases. Sudan enjoys modest gains in household income (0.3–0.4 per cent) and consumption expenditures (0.3–0.4 per cent) during the GERD impounding stage, which improve substantially during the GERD operation phase to 0.9 per cent and 0.8 per cent for household income and household consumption expenditures, respectively. The

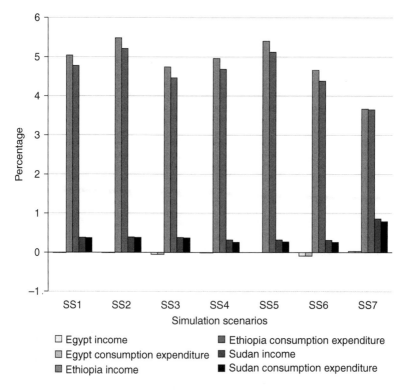

Figure 8.3 Percentage change in household income and consumption expenditures relative to the baseline scenario across the three Blue Nile Basin countries.

GERD impounding and operation have a negligible effect on income and consumption expenditures in Egypt. Egypt experiences a 0.03 per cent decline in household income and consumption expenditure during the GERD impounding and gains 0.03 per cent in income and consumption expenditures when the GERD starts its operation.

The overall welfare effect of the GERD, as measured by the equivalent variation, is substantial (Figure 8.4). Equivalent variation refers to the amount of income that would have to be given to an economy before building the dam so as to leave the economy as well off as it would be after the dam has been built. The GERD impounding induces, on average, a basin-wide welfare gain of about US$1.21 billion. During this phase, Ethiopia and Sudan enjoy a welfare gain of US$1.07 billion and US$192 million, respectively, while Egypt incurs a welfare loss of about US$51 million. The total welfare gain increases to US$1.66 billion when the GERD enters its operation phase. This equals 0.5 per cent of the estimated combined GDP of the Eastern Nile economies in the baseline scenario. All the Eastern Nile countries benefit from the welfare gain due to the GERD operation, although the distribution remains uneven with Ethiopia, Sudan and Egypt earning 66 per cent, 27 per cent and 7 per cent of the total welfare gain, respectively. As the results based on a welfare decomposition analysis (Huff and Hertel 2000) reveal, the endowment effect (i.e. increase in water supply due to the GERD operation), improvements in the commodity terms of trade and investment savings contribute most to the welfare gain in Egypt.

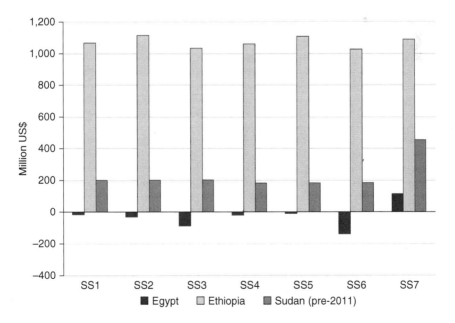

Figure 8.4 Expected welfare effects of the GERD impounding and operation in the Eastern Nile Basin countries.

Welfare gains in Ethiopia and Sudan emanate mainly from the endowment effect (i.e. increase in both water supply and built infrastructure), improved allocative efficiency and technical change in irrigation efficiency.

The increase in Ethiopia's hydropower supply due to the GERD operation influences the patterns of trade in the country. The simulation results demonstrate that exports of hydroelectricity increase by US$522 million, whereas Ethiopia's agricultural exports, which constitute traditionally its main export commodity, decline by US$165 million. Exports of manufactured products also increase by US$54 million. Thus, the GERD operation poses only a marginal risk of 'Dutch disease' for Ethiopia.

Sensitivity analysis

The sensitivity analysis results with respect to the imposed shocks and substitution elasticities of the production structure of the model testify the robustness of the estimated effects of the GERD in the Eastern Nile Basin economies. The sensitivity of the results is tested in the former case by assuming a plus and minus 25 per cent change in hydropower, irrigated land and water supply due to the GERD operation. In the latter case, the applied elasticities of substitution between irrigated land and irrigation water are assumed to be 25 per cent lower and higher. The 95 per cent confidence intervals around the estimated model results are derived following the procedure outlined in Burfisher (2011). The results are presented in Table 8.6 and show no change in sign for the rate of economic growth for the Eastern Nile countries due to the GERD operation. The modelled scenario outcomes presented before fall within the 95 per cent confidence intervals, indicating that these results are robust. The same applies to the changes in the parameter values of the substitution elasticity. Sensitivity analysis tests conducted for the GERD impounding results under average hydrological conditions reveal similar results. The same applies for other variables such as the estimated welfare, income and consumption effects. The model results for both the impounding and operation phases of the GERD are thus robust for a wide range of estimates of the transboundary benefits of the dam.

Table 8.6 Results of the sensitivity analysis

	Original model value	25% change in elasticities of substitution		25% change in hydropower, irrigated land and water	
		Mean % change in real GDP	St.dev.	Mean % change in real GDP	St.dev.
Egypt	0.03	0.03	0	0.03	0
Ethiopia	4.73	4.73	0	4.42–5.04	0.07
Sudan (pre-2011)	1.04	1.04	0	1.00–1.08	0.01

Discussion and conclusions

Employing a multi-region, multi-sector computable general equilibrium framework, this study estimates the economic impacts of the GERD on the Eastern Nile economies. The study evaluates the impact of the dam in a novel fashion taking into account both the transient GERD impounding stage and the long-term GERD operation phase in a global CGE setting. The results of the analysis demonstrate the significance of the GERD in generating basin-wide economic benefits and improving welfare in the Eastern Nile Basin. During the impounding phase, the GERD benefits mainly Ethiopia and to some extent Sudan. For Egypt, the GERD impounding inflicts an economic cost, but this cost is not considered substantial compared to the size of Egypt's economy and the basin-wide benefits the dam is expected to generate. Egypt's economic loss is more pronounced in the less likely case of the GERD impounding occurring in a series of dry years during which Sudan increases its agricultural water withdrawal to fully utilise its quota according to the 1959 agreement with Egypt at the same time. The GERD operation reverses the negative effects of the impounding phase on Egypt's economy and generates substantially higher economic benefits and enhances economic growth and welfare in all the Eastern Nile countries, although the distribution favours especially Ethiopia. The GERD would also contribute to poverty alleviation in the Eastern Nile Basin by increasing real income to unskilled labour.

Our results are more or less consistent with previous findings (Strzepek *et al.* 2008; Aydin 2010), and also concur with the findings of several hydrological models developed to evaluate the downstream impacts of Ethiopia's development of the Blue Nile waters to maximise hydropower production (e.g. Guariso and Whittington 1987; Whittington *et al.* 2005; Blackmore and Whittington 2008). The findings of the present study, like previous studies, reveal that hydropower dams enhance economic growth. Contrary to the findings presented in Ferrari *et al.* (2013), we do not find indications that the GERD would increase the risk of 'Dutch disease' for Ethiopia.

The static nature of the model constitutes a major limitation of the study. A dynamic model would, theoretically, yield more reliable results through the annual adjustments that could be made in national capital stocks in the Eastern Nile countries. However, despite the static nature of our model, the results of the current analysis demonstrate that the role of the GERD in generating basin-wide economic benefits in the Eastern Nile Basin is substantial. In case it occurs during a sequence of dry years, it is worthwhile to extend the impounding period of the dam so as to mitigate the adverse effect it is expected to have on Egypt's economy. However, the findings of the study disclose that Egypt's economy is constrained more by energy than water. Hence instituting a basin-wide power trade scheme that would enable Egypt to import part of the hydropower production expected from the GERD would substantially boost Egypt's economy and thereby further increase the basin-wide economic value of the dam. Issues related to the potential impact of climate change on the development of the

GERD and hence its transboundary economic impacts are deferred for follow-up study on the dam.

References

Aguiar, A., Narayanan, B. and McDougall, R. (2016). An Overview of the GTAP 9 Data Base. *Journal of Global Economic Analysis, 1*(1), 181–208.

Arjoon, D., Mohammed, Y., Goor, Q. and Tilmant, A. (2014). Hydro-Economic Risk Assessment in the Eastern Nile River Basin. *Water Resources and Economics,* 8, 16–31.

Asegdew, G.M., Semu, A.M. and Yosif, I. (2014) Impact and Benefit Study of Grand Ethiopian Renaissance Dam (GERD) during Impounding and Operation Phases on Downstream Structure in the Eastern Nile. In M.M. Assefa, A. Wossenu and G.S. Shimelis (eds), *Nile River Basin: Ecohydrological Challenges, Climate Change and Hydro-politics.* Switzerland: Springer.

Asegdew, G.M. and Semu, A.M. (2014). Assessment of the Impact of the Grand Ethiopian Renaissance Dam on the Performance of the High Aswan Dam. *Journal of Water Resource and Protection,* 6, 583–598.

Aydin, L. (2010). The Economic and Environmental Impacts of Constructing Hydro Power Plants in Turkey: A Dynamic CGE Analysis (2004–2020). *Natural Resources, 1,* 69–79.

Blackmore, D. and Whittington, D. (2008). *Opportunities for Cooperative Water Resources Development on the Eastern Nile: Risks and Rewards.* An Independent Report of the Scoping Study Team to the Eastern Nile Council of Ministers.

Block, P. and Strzepek, K. (2010). Economic Analysis of Large-Scale Upstream River Basin Development on the Blue Nile in Ethiopia Considering Transient Conditions, Climate Variability, and Climate Change. *Journal of Water Resources Planning and Management, 136*(2), 156–166.

Brockmeier, M. (2001). A Graphical Exposition of the GTAP Model. GTAP Technical Paper No. 8 (revised).

Brouwer, R., Hofkes, M. and Linderhof, V. (2008). General Equilibrium Modelling of the Direct and Indirect Economic Impacts of Water Quality Improvements in the Netherlands at National and River Basin Scale. *Ecological Economics,* 66(1), 127–140.

Burfisher, M.E. (2011). *Introduction to Computable General Equilibrium Models.* Cambridge: Cambridge University Press.

Burniaux, J.M. and Truong, T.P. (2002). *GTAP-E: An Energy Environmental Version of the GTAP Model.* GTAP Technical paper No. 16.

Calzadilla, A., Rehdanz, K. and Tol, R.S.J. (2010). The Economic Impact of More Sustainable Water Use in Agriculture: A Computable General Equilibrium Analysis'. *Journal of Hydrology,* 384(3–4), 292–305.

Calzadilla, A., Rehdanz, K. and Tol, R.S.J. (2011). *The GTAP-W Model: Accounting for Water Use in Agriculture.* Working Paper No. 1745, Kiel Institute for the World Economy.

Diao, X. and Roe, T. (2003). Can a Water Market Aver the 'Double-Whammy' of Trade Reform and Lead to a 'Win–Win' Outcome?. *Journal of Environmental Economics and Management, 45*(3), 708–723.

Diao, X., Roe, T. and Doukkali, R. (2005). Economy-Wide Gains from Decentralized Water Allocation in a Spatially Heterogeneous Agricultural Economy. *Environment and Development Economics,* 10, 249–269.

ENTRO (2006). Flood Preparedness and Early Warning. Technical Background Paper. Addis Ababa, Ethiopia.

ENTRO (2009). Eastern Nile Irrigation and Drainage Studies Cooperative Regional Assessment: Analysis Report. Addis Ababa, Ethiopia.

FAO (2005). Irrigation in Africa in Figures: AQUASTAT Survey 2005. Rome, Italy.

FAO (2013). FAO Statistical Yearbook. Rome, Italy.

Feng, S., Li, L.X., Duan, Z.G. and Zhang, J.L. (2007). Assessing the Impacts of South-to-North a Water Transfer Project with Decision Support Systems. *Decision Support Systems in Emerging Economies, 42*(4), 1989–2003.

Ferrari, E., McDonald, S. and Osman, R. (2013). Grand Ethiopian Renaissance Dam: A Global CGE Model to Assess the Economic Effects on the Ethiopian Economy. Paper prepared for the 16th Annual Conference on Global Economic Analysis, 'New Challenges for Global Trade in a Rapidly Changing World', Shanghai, China.

Gomez, C.M., Tirado, D. and Rey-Maquieira, J. (2004). Water Exchanges Versus Water Work: Insights from a Computable General Equilibrium Model for the Balearic Islands. *Water Resources Research, 40*(10), 1–11.

Goor, Q., Halleux, C., Mohamed, Y. and Tilmant, A. (2010). Optimal Operation of a Multipurpose Multireservoir System in the Eastern Nile River Basin. *Hydrology and Earth System Sciences, 14*(10), 1895–1908.

Guariso, G. and Whittington, D. (1987). Implications of Ethiopian Water Development for Egypt and Sudan. *International Journal of Water Resource Development, 3*(2), 105–114.

Hertel, T.W. (1997) *Global Trade Analysis: Modeling and Applications.* Cambridge: Cambridge University Press,

Horridge, J.M. (2011a). *SplitREG: A Program to Create a New Region in a GTAP Database.* Centre of Policy Studies, Monash University, Melbourne, Australia.

Horridge, J.M. (2011b). *GTAPAdjust: A Program to Balance or Adjust a GTAP Database.* Centre of Policy Studies, Monash University, Melbourne, Australia.

Huff, K.M. and Hertel, T.W. (2000). *Decomposing Welfare Changes in the GTAP Model.* GTAP Technical Paper No. 5.

ILO (2013). *Decent Work Country Profile Ethiopia.* ILO Country Office for Ethiopia and Somalia. Addis Ababa, Ethiopia.

Jeuland, M. (2010a). Economic Implications of Climate Change for Infrastructure Planning in Transboundary Water Systems: An Example from the Blue Nile'. *Water Resources Research, 46*(11), W11556. Doi: 10.1029/2010WR009428.

Jeuland, M. (2010b). Social Discounting of Large Dams with Climate change Uncertainty. *Water Alternatives, 3*(2), 185–206.

Kahsay, T.N., Kuik, O., Brouwer, R. and van der Zaag, P. (2015). Estimation of the Transboundary Economic Impacts of the Grand Ethiopian Renaissance Dam: A Computable General Equilibrium Analysis. *Water Resources and Economics, 10,* 14–30.

MDI consulting engineers (2012). The Grand Ethiopian Renaissance Dam Project Report: Initial Transboundary Environmental Impact Assessment. EEPCo, Addis Ababa, Ethiopia.

MoFA (2013). A Week in the Horn of Africa: The International Panel of Experts. Report on the Grand Ethiopian Renaissance Dam. Retrieved 3 January 2014, from www.mfa.gov.et/weekHornAfrica/morewha.php?wi=1024#1026,.

MoFED (2010). Growth and Transformation Plan 2010/11–2014/15. Addis Ababa, Ethiopia.

Nelson, G.C., Rosegrant, M.W., Palazzo, A., Gray, I., Ingersoll, C., Robertson, R., Tokgoz, S., Zhu, T., Sulser, T.B., Ringler, C., Msangi, S. and You, I. (2010). Food

Security, Farming, and Climate Change to 2050: Scenarios, Results, Policy Options. Washington, DC: International Food Policy Research Institute.

Peters, J.C. (2016). 'The GTAP-Power Data Base: Disaggregating the Electricity Sector in the GTAP Data Base'. *Journal of Global Economic Analysis, 1*(1), 209–250.

Robinson, S., Strzepek, K., El-Said, M. and Lofgren, H. (2008). The High Dam at Aswan. In R. Bhatia, R. Cestti, M. Scatasta and R.P.S. Malik (eds), *Indirect Economic Impacts of Dams: Case Studies from India, Egypt and Brazil*. Washington, DC: The World Bank.

Seung, C.K., Harris, T.R., Englin, J.E. and Netusil, N.R. (2000). Impacts of Water Reallocation: A Combined Computable General Equilibrium and Recreation Demand Model Approach. *Annals of Regional Science, 34*(4), 473–487.

Seung, C.K., Harris, T.R., MacDiarmid, T.R. and Shaw, W.D. (1998). Economic Impacts of Water Reallocation: A CGE Analysis for the Walker River Basin of Nevada and California. *Journal of Regional Analysis and Policy, 28*(2), 13–34.

Strzepek, K.M., Yohe, G.W., Tol, R.S.J and Rosegrant, M.w. (2008). The Value of the High Aswan Dam to the Egyptian Economy. *Ecological Economics, 66*, 117–126.

Tesfaye, A., Wolanios, N. and Brouwer, R. (2016). Estimation of the Economic Value of the Ecosystem Services Provided by the Blue Nile Basin in Ethiopia. *Ecosystem Services, 17*, 268–277.

Tractebel, Coyen et Bellier and GDF Suez (2011). *Hydrological and Reservoir Simulations Studies, GERD Project Impounding and Operation Simulations Impact Study on High Aswan Dam*. Addis Ababa: EEPCo.

Van Heerden, J.H., Blignaut, J. and Horridge, M. (2008). Integrated Water and Economic Modeling of the Impacts of Water Market Instruments on the South African Economy. *Ecological Economics, 66*(1), 105–116.

Whittington, D., Wu, X. and Sadoff, C. (2005). Water Resources Management in the Nile Basin: The Economic Value of Cooperation. *Water Policy, 7*, 227–252.

Wittwer, G. (2009). The Economic Impacts of a New Dam in South-East Queensland. *The Australian Economic Review, 42*(1), 12–23.

9 From projecting hydroclimate variability to filling the GERD

Upstream hydropower generation and downstream releases

Ying Zhang, Solomon Tassew Erkyihun and Paul Block

Background and introduction

Located upstream of Blue Nile River in Ethiopia, GERD will be the largest hydroelectric power plant in Africa upon completion and the second largest reservoir on the Nile River next to Lake Nasser behind the High Aswan Dam in Egypt. While it is undoubtedly important for Ethiopia's economic development given its large hydropower potential, downstream countries Sudan and Egypt are concerned with the expected streamflow reduction, particularly during the filling stage – a potential threat to lives and livelihood that rely on the Blue Nile River. The tenor of talks between riparian countries has vacillated since 2011, shortly after construction of GERD commenced, ranging from handshakes to military threats (Maher 2013; Gebreluel 2014; Salman 2016). However, no consensus on an agreeable filling policy has yet been reached, highlighting the difficulty in resolving the issue.

This chapter addresses two key aspects across the reservoir filling stage – inflow to the reservoir based on hydroclimatic conditions and subsequently the outflow (releases) based on filling policies – to investigate the trade-off between upstream hydropower generation and downstream releases. In this analysis, releases from GERD are not evaluated further downstream; however, streamflow into Sudan and Egypt can be inferred based on previous studies (e.g. Zhang *et al.* 2015). Adaptive operational policies in Sudan and Egypt in coordination with the GERD's filling policies may modulate downstream impacts and improve overall basin water management. This coordinated operation is not explicitly addressed in here; the reader is referred to Wheeler *et al.* (2016) for details.

In order to construct plausible projections of future streamflow into the reservoir, both inter-annual variability and low-frequency signals (decadal hydroclimate oscillations) in historical streamflow are identified. This mimics patterns in precipitation over the upper Blue Nile Basin, which demonstrates strong intra- and inter-annual variability. The annual precipitation cycle peaks during the Kiremt (main) rainy season, spanning June–September, which contributes approximately 70 per cent of annual basin-wide precipitation on average. Dry conditions in the basin are often associated with the El Niño phenomenon, occurring, on average, every three–eight years (e.g. Bekele 1997;

Wolde-Georgis 1997; Block and Rajagopalan 2007; Diro *et al.* 2011). Low-frequency oscillations can lead to multi-decadal variability, where a general above (or below) normal precipitation condition is latent across extended periods.

Evaluation of low-frequency signals in precipitation and streamflow in the upper Blue Nile Basin has received relatively little attention, given that it may explain a small fraction of the total variance. However, it is important when considering filling policies, expected hydropower generation and downstream releases, as they are closely linked to reservoir inflow conditions across the filling phase. For instance, anomalously high flow conditions across filling years could produce drastically different outcomes than if low flow conditions persist. Thus, investigation into the presence of low-frequency signals in streamflow is warranted.

Projecting hydroclimate variability

To evaluate the existence of low-frequency signals in average basin-wide precipitation over the Blue Nile Basin, wavelet analysis is performed based on the CRU TS v3.23 gridded global monthly precipitation dataset from 1901 to 2014 (Harris *et al.* 2014). Wavelet analysis provides a mechanism to investigate the precipitation in time and frequency domains simultaneously, such that significant modes of variabilities (signals) at certain frequencies and times can be identified. A Morlet mother wavelet is chosen for its properties suitable for precipitation time series analysis (Kwon *et al.* 2007). For a more detailed description of wavelet analysis, see Kumar and Foufoula-Georgiou (1997) and Torrence and Compo (1998). The precipitation time series is aggregated into annual values and subsequently detrended and normalised for wavelet analysis. The annual sequence helps to reduce seasonality effects and potentially emphasise any low-frequency signals. Consequently, two low-frequency bands that are significantly different from the background noise spectrum (i.e. white noise) are identified – one at a twenty-one-year (90 per cent significance) period and the other at thirty-five-year (95 per cent significance) period.

Given the presence of both interannual and low-frequency variabilities, inflow projections across the filling period are conditioned on both, as described below. Projections start in 2015, the end of the precipitation dataset, and continue to 2034. Filling of the GERD reservoir, however, is not expected to commence until 2018. Accordingly, filling spans for 204 months (2018–2034), while the full simulation period is 240 months (2015–2034). Beginning with the low-frequency periods, time series associated with each significant peak can be reconstructed (Torrence and Compo 1998) and then together scaled to the historical period.

The scaled time series are projected beyond 2015 independently using a non-parametric Block-KNN method (Erkyihun *et al.* 2016), and subsequently summed. The general methodology is illustrated in Figure 9.1 using the twenty-one-year low-frequency signal as an example. More details can be found in Zhang *et al.* (2016).

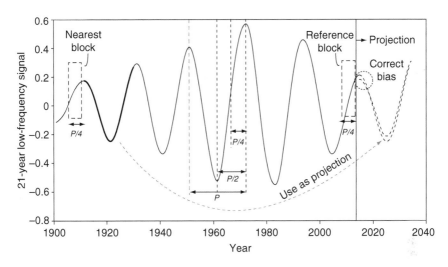

Figure 9.1 Illustration of the Block-KNN projection methodology.
Source: Zhang *et al.* 2016.

For this chapter, a single projection is proposed for each of the two significant low-frequency signals identified, given the limited amount of variance they explain (17 per cent). Alternatively, the uncertainty associated with the low-frequency projections could be accounted for by repeating the Block-KNN method many times and selecting from the full range of nearest neighbours (not simply the closest) to create a projection envelope.

In contrast, monthly time series with the low-frequency signal removed are projected stochastically to represent the expected range of precipitation within each month. The low-frequency signal is assumed to contribute equivalently across all months in a year; that is, seasonality is preserved in the residual monthly time series. Figure 9.2 illustrates the aggregated low-frequency signals and annual and monthly residuals across the historical record.

To construct the projections (2015–2034), twenty years are randomly bootstrapped with replacement from the historical period (1901–2014); for each year, the twelve monthly residuals are added to the projected aggregated low-frequency value to form monthly projections with the low-frequency signal embedded (Figure 9.3 illustrates one simulation).

Corresponding monthly temperature time series projections are also needed in addition to the precipitation projections, to generate streamflow. To preserve the relationship between precipitation and temperature, for each month of the precipitation projections, the month with the most similar precipitation value (in the same month, e.g. Januaries only) from the historical (CRU TSv3.32) record is selected and both precipitation and temperature from that month are retained. These, then, form the final set of projections.

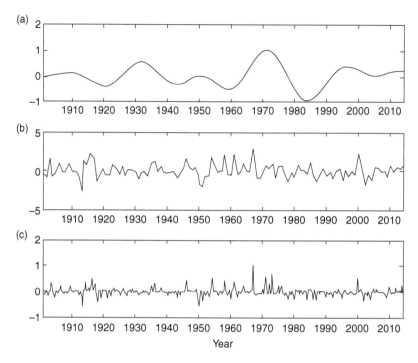

Figure 9.2 (a) Aggregated low-frequency signal in Blue Nile annual precipitation over 1901–2014; values are scaled. (b) Residual annual time series after subtracting the signal in (a) from the detrended normalised historical record. (c) Residual monthly time series after disaggregating from annual time series to monthly time series and subtracting the signal in (a).

Source: Zhang *et al.* 2016.

To translate the precipitation projection into monthly streamflow entering the GERD reservoir, a variation of the WatBal water balance model (Yates 1996) is calibrated and adopted. The model has been shown to satisfactorily reproduce monthly streamflow observations along the Nile River given precipitation and other inputs (Yates and Strzepek 1998; Zhang *et al.* 2015). Three modules are included in the model: a soil moisture modelling scheme to account for water fluxes, potential evapotranspiration using the Hargreaves method (Hargreaves and Samani 1982), and a storage scheme for lakes and swamps. Additional details are given by Yates (1996) and Zhang *et al.* (2015).

Filling the GERD

Linked with the WatBal model, which gives the projected inflows, the GERD dam and reservoir model then simulates monthly storage volume (time to fill the reservoir), hydropower generation and releases for downstream countries, contingent on the filling policy selected. The model is run for twenty years, representing 2015–2034. Dam and reservoir design characteristics are developed from publicly available sources, the International Panel of Experts report (Elsayed *et al.* 2013) and a preliminary study conducted by the United States Bureau of Reclamation (USBR 1964). Original plans allow for up to 9 km³ of storage by 2015 with two 375 MW turbines online to generate hydropower (EEPCo 2013). Upon dam completion in the latest goal of 2018, further storage may continue up to the full supply capacity of 74 km³ and all turbines will be available for generating hydropower up to a rated capacity of 6,000 MW. Therefore, in this study, filling is modelled to start in January 2018 with an initial storage of 9 km³ considered.

Filling policy specifics

The rate at which the reservoir is filled has direct implications for hydropower generation and downstream releases. Three types of filling policies are considered here, including fractional, threshold and absolute.

Fractional filling policies allow impounding a specified percentage of total monthly streamflow into the GERD reservoir; both 10 per cent and 25 per cent are evaluated here. This policy guarantees that some quantity of water can be impounded, although this quantity varies month to month and year to year (King and Block 2014). Fractional policies generally favour a sharing by upstream and downstream countries of the risk associated with streamflow variability. In this case, the general water balance is:

$$V_t = V_{t-1} + X\% * Q_t - E_t \qquad\qquad \text{(Equation 1)}$$

where V is reservoir volume, t is time (month), X *per cent* is the fraction impounded (here, either 10 per cent or 25 per cent), Q is streamflow volume into the reservoir and E is net evaporation volume from the reservoir.

The threshold filling policy allows any streamflow volume in excess of the long-term historical monthly average to be impounded in the reservoir. Thus, in months with anomalously high flow, the volume impounded can be large; however, in months with flow below the average, no impounding is permitted (King and Block 2014). In months with flow below the average, Ethiopia is not required to make up the difference with existing reservoir storage. This policy generally favours downstream countries, as water for impoundment in any given month is not guaranteed, and downstream countries receive at least long-term average monthly streamflow or the full flow volume in drier than average months. In this case, the general water balance is:

$$V_t = V_{t-1} + \max(Q_t - HASF_t, 0) - E_t \qquad\qquad \text{(Equation 2)}$$

where $HASF$ is the long-term historical average streamflow volume into the reservoir.

Absolute filling policies allow for a guaranteed volume of water to be impounded in the reservoir annually. These policies are structured here based on time taken to fill the reservoir, namely four, six, or eight years. To fill the reservoir in four years, for example, one-fourth of the total reservoir volume may be impounded each year, irrespective of the streamflow. Annual flow impounded is disaggregated to monthly flow weighted by the long-term historical monthly average, and then adjusted by a factor γ such that the expected time to fill (e.g. four, six, or eight years) can be guaranteed (minus the effect of evaporative loss). It is possible, therefore, that in abnormally low flow months, insufficient water will be available to meet the full impoundment demand, and all water will be stored with no releases downstream. These policies generally favour upstream countries, as water for impoundment in any given month is guaranteed and the expected time to fill the reservoir is fixed. In this case, the general water balance is:

$$V_t = V_{t-1} + \gamma * U_t - E_t \qquad \text{(Equation 3)}$$

where U is the weighted monthly impoundment volume and t is in months.

Thus, six filling policies (of three general types) are tested. It is important to note that all water not impounded may be passed through turbines to generate hydropower.

Upstream hydroelectric generation

Given the 9 km³ allowable reservoir storage volume prior to the start of the official filling stage (January 2018), only an additional 5.8 km³ of storage is required to reach the minimum operation level from which hydropower generation may commence at full speed (Elsayed et al. 2013). For most filling policies, this occurs in the first year of filling; annual hydroelectric generation then increases rapidly through the filling stage. Annual hydroelectric generation varies from 9,000 GWh to 10,400 GWh, with policies releasing more water (storing less) typically generating more energy (Table 9.1). All filling policies have simulations in which they

Table 9.1 Annual hydroelectric generation (GWh) during the filling stage (unique to each policy) and full projection period (2018–2034), assuming full hydroelectric generation after the reservoir is filled (15,000 GW-hrs annually). Calculations are based on the mean values of 100 simulations and numbers are rounded to hundreds

Filling policy		10%	25%	HASF	Four-year	Six-year	Eight-year
During filling stage	Mean	10,400	9,800	9,300	9,000	9,800	10,100
	Max.	11,500	11,000	11,000	10,000	11,000	11,200
2018–2034	Mean	11,000	13,200	13,000	13,600	13,200	12,700

Source: Zhang et al. 2016.

generate 11,000 GWh or more, averaged over the filling stage, except for the four-year policy, which has a maximum of 10,000 GWh across the filling stage (Table 9.1). This is not surprising, as the four-year policy emphasises storing water and releasing less downstream flows. However, because the four-year policy fills the most quickly, the ensuing years are likely to produce more hydropower than under other policies. For example, the 13,600 GWh produced, on average, under the four-year policy across 2018–2034 surpasses all other policies. (This result should be used primarily for comparison with other policies and not in terms of absolute values, given the aforementioned discussion regarding operational policies after filling. For the calculations here, once the reservoir is filled, generation is expected to be approximately 15,000 GWh/year [Elsayed *et al.* 2013].)

It is also important to consider the discounting effect on hydroelectricity benefits, that is, the hydropower generated early will be more valuable than that in the future. Assuming that electricity would be sold for US$0.07/kWh (ENTRO 2007) and adopting a 3 per cent annual discounting rate, the results (Table 9.2) indicate that during the filling stage, the 10 per cent filling policy produces a discounted present value on the order of US$528 million. This is substantially lower than the annual benefits resulting from the 25 per cent, six-year and eight-year policies across the filling stage, which are all approximately equal (~US$575 million).

The corresponding time required to fill the reservoir to the full supply level under each filling policy is also reported (Table 9.3). Based on the 100 simulations, the four-year filling policy, not surprisingly, tends to fill the most quickly, with a median filling time of only forty-seven months and 95 per cent of the simulations filling within fifty-four months. The median time to fill is approximately equivalent for the 25 per cent (sixty-eight months), HASF (sixty-eight months) and six-year (seventy-one months) filling policies. However, considering the time to fill associated with the lower 95th percentile based on the simulations, while the 25 per cent and six-year policies differ by only eight–twelve months from their respective medians, the HASF policy requires more than 204 months, implying that the reservoir would not fill even by 2034. Thus the 25

Table 9.2 Discounted present value (as of December 2014) of annual benefit (US$ million) from hydroelectricity generated during the filling stage (unique to each policy) and full projection period (2018–2034), assuming electric price of $0.07/kWh, 3 per cent annual discounting rate and full hydroelectric generation after the reservoir is filled (15,000 GWh annually). Calculations are based on the mean values of 100 simulations.

Filling policy		10%	25%	HASF	Four-year	Six-year	Eight-year
During filling stage	Mean	527.7	573.8	541.0	553.9	575.2	574.4
	Max.	644.9	619.5	623.5	632.8	638.9	625.1
2018–2034	Mean	535.5	643.9	632.4	668.9	644.4	620.3

Source: Zhang *et al.* 2016.

Table 9.3 Time to fill the reservoir for different filling policies

Filling policy		10%	25%	HASF	Four-year	Six-year	Eight-year
Number of months to fill	Median	178	68	68	47	71	95
	Lower 95th percentile	>204	80	>204	54	79	103
	Median-95th	>26	12	>136	7	8	8
Median year filled		2032	2023	2023	2021	2023	2025

Source: Zhang *et al.* 2016.

per cent and six-year policies have relatively low variance in their time to fill, which is advantageous for reservoir planning; in contrast, the HASF policy has a large variance in the time to fill, under which downstream streamflow would remain relatively close to the historical average each month. The 10 per cent policy clearly requires the longest time to fill the reservoir.

Downstream releases during the filling stage

Mean annual downstream releases from the GERD vary between 33–46 km^3 at the beginning of the filling stage, dependent on the filling policy. The four-year policy clearly has the largest initial abstractions, resulting in significantly reduced flows downstream; however, flows return to normal faster than under any other policies. In contrast, the 10 per cent policy produces relatively stable releases fluctuating around 45 km^3, not returning to normal flow until the end of the projection period. Considering the lower 95th percentile of simulations, releases fall below 15 km^3 under the four-year policy, and even below 30 km^3 under the 10 per cent and HASF filling policies. Here the HASF policy produces the largest releases throughout the filling stage, on average, and also represents the smallest gap relative to median releases. This relatively low uncertainty contrasts with the large HASF policy uncertainty in the time to fill, and emphasises its favourableness to downstream countries.

The percentage reduction in annual downstream releases for each filling policy (each simulation compared with historical averages) is also computed. Unsurprisingly, for the percentage-based filling policies, the percentage reduction during the filling stage is simply the percentage itself (10 per cent and 25 per cent) for each simulation resulting in no variability. For the absolute filling policies, the variability of relative reduction under the four-year policy is higher than under the six- or eight-year policy. Though the HASF policy has the highest variability among all filling policies during the filling stage, the absolute amount of streamflow downstream of the GERD is often simply the long-term historical average; for this policy, reductions imply that water was impounded during a wet year. Finally, the median relative reductions for the four-, six- and eight-year policies are approximately 35 per cent, 24 per cent and 18 per cent, respectively.

Under the absolute filling policies, months with no downstream releases are possible. This occurs during an exceptionally low flow month, when streamflow is less than the allotted impoundment volume. Not surprisingly, these months occur more often under the four-year policy than under the six- or eight-year policy. For example, only nine no flow months occur under the eight-year policy, summing across all 100 simulations; while this is minimal, it is by no means trivial. The no release months occur most frequently in February and March, the dry season in the Blue Nile basin (Figure 9.4). While under the four-year policy, some no flow months occur even during the high flow season (September–October) in a few simulations. Thus, while the absolute filling policies have small uncertainty in the time to fill, the possibility of months with no downstream flow may pose a serious threat to downstream countries.

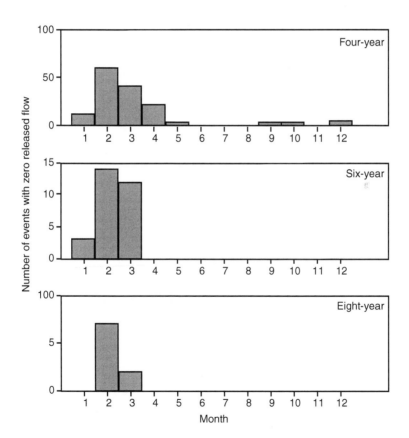

Figure 9.4 Number of months with no downstream releases for the three absolute filling policies, accumulated from the 100 simulations.

Source: Zhang *et al.* 2016.

Note
The y-axis scale differences for each subplot.

Discussion

No agreeable filling policy for the reservoir behind the GERD has been established among the riparian countries, even though serious upstream and downstream impacts are at stake. Two major factors – projected streamflow and various filling policies – are explored here.

Inclusion of the low-frequency signals is expected to produce more plausible projections across the filling period. The uncertainty in these low-frequency signals, however, has not been explicitly included; rather, a static approach is undertaken. Additional investigation into the effects on outcomes of interest of stochastically representing the low-frequency signals in projections may be worthwhile.

Regarding filling, the 10 per cent policy is probably impractical given the excessively long time required to fill the reservoir. The four-year absolute policy lies at the other end of the spectrum, with a rapid filling time, implying sharp reductions downstream and likely occurrences of no flow months; therefore, it is unlikely to be selected if the decision is to be mutual. The HASF policy typically favours downstream countries regarding streamflow, but results in significant uncertainty in the time to fill. The three remaining policies (25 per cent, six-year and eight-year) fall somewhere in the middle and may be considered as compromise solutions. The 25 per cent and six-year policies are nearly identical in terms of time to fill and hydroelectricity generated; however, the 25 per cent policy has a larger uncertainty in outcomes while the six-year policy suffers from the possibility of months with no flow releases. Thus, a hybrid combination may be warranted, combining the percentage-based and absolute filling policies to reduce uncertainty and eliminate no-flow months, or alternatively, selecting a minimum required flow to be released per month. The goal here is not to be exhaustive but rather to be illustrative in how different filling policies compare and whether they may favour upstream or downstream countries. Consensus among the riparian countries prior to initiating filling is strongly encouraged.

Note that gridded data are used here in lieu of station observations due to their longer record length, which is desirable for low-frequency signal analysis. Gridded data can cause overestimation of the low flows and underestimation of the high flows, even with a carefully calibrated WatBal model. However, the cumulative calibration error in terms of annual volume is approximately – 750 million m^3, which is minimal compared to the substantial natural inflow (approximately 46,000 million m^3), having a relatively minor effect when comparing across filling policies.

As discussed earlier, no attempt is made to include post-filling operational policies. This will clearly be important for infrastructure management in all three countries. Sudan may be the least affected, given its relatively small amount of current storage capacity. Coordination of the GERD and the Aswan Dam in Egypt, however, is crucial for maximisation of regional net benefits. In fact, cooperation among the three riparian countries at all stages – construction, filling and long-term management – is critical for project success and future regional development.

References

Bekele, F. (1997). Ethiopian Use of ENSO Information in Its Seasonal Forecasts. *Internet Journal of African Studies*.

Block, P.J. and Rajagopalan, B. (2007). Interannual Variability and Ensemble Forecast of Upper Blue Nile Basin Kiremt Season Precipitation. *Journal of Hydrometeorology*, 8, 327–343.

Diro, G.T., Grimes, D.I.F. and Black, E. (2011). Teleconnections between Ethiopian Summer Rainfall and Sea Surface Temperature: Part I – Observation and Modelling. *Climate Dynamics*, 37, 103–119.

EEPCO (2013). *Grand Ethiopian Renaissance Dam*. Addis Ababa: Ethiopian Electric Power Corporation.

Elsayed, S.M., Hamed, K., Asfaw, G., Seleshi, Y., Ahmed, A.E., Deyab, D.H., Yon, B., Roe, J.D.M., Failer, E. and Basson, T. (2013). *International Panel of Experts for Grand Ethiopian Renaissance Dam*. Addis Ababa: IPoE.

ENTRO (2007). Coordinated Investment Planning – Generation. *Eastern Nile Power Trade Program Study*. AfDB.

Erkyihun, S.T., Rajagopalan, B., Zagona, E., Lall, U. and Nowak, K. (2016). Wavelet-Based Time Series Bootstrap Model for Multidecadal Streamflow Simulation using Climate Indicators. *Water Resources Research*, 52, 4061–4077.

Gebreluel, G. (2014). Ethiopia's Grand Renaissance Dam: Ending Africa's Oldest Geopolitical Rivalry?. *The Washington Quarterly*, 37, 25–37.

Hargreaves, G.H. and Samani, Z.A. (1982). Estimating Potential Evapotranspiration. *Journal of the Irrigation and Drainage Division*, 108, 225–230.

Harris, I., Jones, P.D., Osborn, T J. and Lister, D.H. (2014). Updated High-Resolution Grids of Monthly Climatic Observations – The CRU TS3.10 Dataset. *International Journal of Climatology*, 34, 623–642.

King, A. and Block, P. (2014). An Assessment of Reservoir Filling Policies for the Grand Ethiopian Renaissance Dam. *Journal of Water and Climate Change*, 5(2), 233–243.

Kumar, P. and Foufoula-Georgiou, E. (1997). Wavelet Analysis for Geophysical Applications. *Reviews of Geophysics*, 35, 385–412.

Kwon, H.-H., Lall, U. and Khalil, A.F. (2007). Stochastic Simulation Model for Nonstationary Time Series using an Autoregressive Wavelet Decomposition: Applications to Rainfall and Temperature. *Water Resources Research*, 43, n/a-n/a.

Maher, A. (2013). *Egyptian Politicians Caught in On-Air Ethiopia Dam Gaffe*. BBC News. Retrieved 25 January 2016, from www.bbc.com/news/world-africa-22771563.

Salman, S.M.A. (2016). *Sudan: Renaissance Dam – Limelight On the Fourth Ministerial Meeting & Khartoum Document*. Khartoum: SudaNow. Retrieved 25 January 2016, from http://allafrica.com/stories/201601080844.html.

Torrence, C. and Compo, G.P. (1998). A Practical Guide to Wavelet Analysis. *Bulletin of the American Meteorological Society*, 79, 61–78.

USBR (1964). Land and Water Resources of the Blue Nile Basin: Ethiopia. In Reclamation, U.S.B.O. & U.S. Dept. of the Interior, G. (eds). Washington, DC: Main Rep. and Appendices.

Wheeler, K.G., Basheer, M., Mekonnen, Z.T., Eltoum, S.O., Mersha, A., Abdo, G.M., Zagona, E.A., Hall, J.W. and Dadson, S.J. (2016). Cooperative Filling Approaches for the Grand Ethiopian Renaissance Dam. *Water International*, 41(4), 611–634.

Wolde-Georgis, T. (1997). El Niño and drought early warning in Ethiopia. *Internet J. Afr. Studies*, 2.

Yates, D.N. (1996). WatBal: An Integrated Water Balance Model for Climate Impact Assessment of River Basin Runoff. *International Journal of Water Resources Development, 12*, 121–140.

Yates, D.N. and Strzepek, K.M. (1998). Modeling the Nile Basin under Climate Change. *Journal of Hydrologic Engineering, 3*(2), 98–108.

Zhang, Y., Block, P., Hammond, M. and King, A. (2015). Ethiopia's Grand Renaissance Dam: Implications for Downstream Riparian Countries. *Journal of Water Resources Planning and Management, 141*(9), 05015002.

Zhang, Y., Erkyihun, S.T. and Block, P. (2016). Filling the GERD: Evaluating Hydroclimatic Variability and Impoundment Strategies for Blue Nile Riparian Countries. *Water International, 41*, 593–610.

10 Managing risks while filling the Grand Ethiopian Renaissance Dam

Kevin G. Wheeler

Introduction

The Grand Ethiopian Renaissance Dam (GERD) offers a timely opportunity for cooperation among Ethiopia, Sudan and Egypt. The potential electrification benefits of the GERD are clear to Ethiopia and any countries connected through transmission lines (EDF and Scott Wilson 2007), however, there remains concerns from downstream countries regarding how the GERD will affect the availability of water for irrigation, municipal, environmental and power generation within their borders. This concern has been the focus of numerous debates among stakeholders and institutions within the geopolitically complex region. This study explicitly does not attempt to address the issues of this larger context, but focuses exclusively on the physical characteristics of the system including descriptions of the known operations of current infrastructure, along with the potential for increased coordination and collaboration among the parties involved as the GERD becomes operational.

An International Panel of Experts commissioned by the three countries indicated the need for further analysis of the period during which the 74 billion cubic metre (BCM) storage reservoir behind the GERD will be initially filled (IPoE 2013). The leaders of the three countries exemplified their willingness to cooperate through a Declaration of Principles (DoP 2015), however, the technical specifications of how cooperation would occur with respect to reservoir filling have yet to be established. This critical period may result in the first effects to downstream countries, however, it also provides the first opportunity to translate the principles of cooperation into tangible actions.

To understand the potential impacts in the Eastern Nile Basin during the filling period of the GERD, numerous filling strategies were developed from combinations of:

1   various operations of the GERD during filling
2   starting conditions of the High Aswan Dam (HAD)
3   modifications to the current operations (reoperation) of the Sudanese and Egyptian Reservoirs
4   explicit coordination of releases from the GERD to avoid critical downstream impacts.

A risk-based framework was used to evaluate the filling strategies and scenarios using 103 sequences of hydrologic inflow data derived from a reconstructed historical record. Many potential solutions were identified that could be considered during the negotiations, yet we acknowledge and emphasise the complexities of implementing any cooperative strategies. For example, challenges are inherent in establishing data sharing procedures, quality assurance protocols and designing conflict resolution mechanisms among participants. Such challenges must not be underestimated and mutual risk reduction depends on these being overcome to allow solutions to move from unilateral management to dynamically cooperative approaches (Sadoff and Grey 2005).

An agreement ultimately should not be based solely on technical studies, however, we argue that a successful negotiation can be supported through a well-designed hydro-policy modelling framework which provides sufficient accuracy, transparency and flexibility for stakeholders to develop and test innovative solutions and explore the trade-offs and benefits of compromise. Such a framework provides a sufficiently detailed representation of the physical characteristics of a water management system alongside an accurate representation of the relevant existing and potential operational policies. This study intentionally does not provide a particular 'optimal' filling solution, but instead suggests potential pathways of mutual benefit through joint management. Further exploration of the possible solutions by stakeholders should be conducted within a wider water diplomacy framework (Islam and Susskind 2012).

Previous modelling studies

Several modelling analyses have been performed over the years to quantify the development potential of the Nile and its tributaries and assess the benefits and risks to downstream countries. Guariso and Whittington (1987) use a multi-objective optimisation framework to demonstrate the alignment of interests between Ethiopian hydropower and Egyptian and Sudanese agriculture. The economic benefits to Ethiopia of large-scale Blue Nile development are shown by Block and Strzepek (2010), however, they identify a reduced degree of benefit when considering the potential effects of climate changes. Similarly, McCartney and Girma (2012) provide an analysis of multi-use infrastructure development within Ethiopia and the resulting benefits to Ethiopian agriculture and hydropower, while considering the risks of climate change and reduced flows. A hydro-economic framework for integrating climate change impacts into infrastructure planning within the Eastern Nile is presented by Jeuland (2010) and similarly shows a high sensitivity of economic benefits to runoff conditions. This work is extended in Jeuland and Whittington (2014) using a real-options approach to analyse the selection, sizing, sequencing and operation of future reservoirs within Ethiopia. Goor et al. (2010) use a stochastic dual dynamic programming approach within a hydro-economic framework to optimise operations of a potential multi-purpose hydropower project in the Blue Nile. Arjoon et al. (2014) reapplied and refined this framework as the construction of the GERD became a reality. Although some of these studies discuss the operation and filling of possible reservoirs, they often contain either simplified or idealised reservoir operations. To be used as a tool for supporting multilateral negotiations, models must simulate the complexity of existing and potential dam operations. Any agreement on new strategies must fit within these political and practical constraints, hence, a robust 'hydro-policy' modelling framework as described earlier is suggested.

Decision support systems (DSS) are another class of computer tools that have been developed for the analysis of the Nile Basin. These are designed to be used by multiple stakeholders and generally commissioned by institutions. Yao and Georgakakos (2003) integrate a water resource model called the Nile Decision Support Tool (DST) with a database to bring together vast amounts of spatially and temporally discrete and distributed hydrologic data. Following the formation of the Nile Basin Initiative (NBI) in 1999, a new user-accessible platform, titled the Nile Basin DSS, was developed to facilitate future technical analysis across the basin by incorporating a variety of models designed for specific analytical purposes while drawing from a common database (NBI 2014). Modelling platforms that focus on the long-term planning and development of water resources such as MIKE HYDRO (Jonker et al. 2012) and WEAP (Yates et al. 2005) were initially integrated into the Nile DSS. Another selection of models of the Eastern Nile, including SWAT (Hassan 2012), Ribasim (van der Krogt and Ogink 2013) and RiverWare (Wheeler and Setzer 2012), was concurrently developed by the Eastern Nile Regional Technical Office (ENTRO) of NBI. These all have the potential to be integrated into the Nile Basin DSS in the future, with the particular value of RiverWare's capability to simulate complex operations of multiple reservoir systems (Zagona et al. 2001).

The filling of the GERD has been considered in four recent studies. Bates et al. (2013) analysed specific fixed monthly release patterns that range from 20.8 to 40.0 BCM/year under *average*, *moderate drought* and *severe drought* scenarios and using three starting elevations of the HAD. A combination of tools was used in this study including MIKE BASIN and the RAPSO model (EDF and Scott Wilson 2007) and separate runs were required to capture the transition of policies from filling to normal operations. Similarly, Mulat and Moges (2014) used MIKE HYDRO to simulate a single historical period of 1973–1978 that represents 'average' conditions to analyse a pre-defined single six-year filling strategy. This study considers a single hydrologic inflow node on each the Blue and White Nile tributaries, includes the GERD and HAD, but contains no information on Sudanese reservoirs or any intervening flows. While these studies provide insight to specific scenarios under specific hydrologic conditions, the deterministic nature of these analyses limits their usefulness to evaluate future risks.

King and Block (2014) and Zhang et al. (2015) develop stochastically generated inflows from a precipitation-driven hydrologic model and apply these to a water resource model that evaluates five potential filling policies. These filling strategies include the retention of 5 per cent, 10 per cent and 25 per cent of inflows, and retention of flows over the historical annual mean average (HASF) and 90 per cent of the HASF. Similarly, Keith *et al.* (2017) use a system dynamics approach to evaluate a wide range of simple filling rates ranging from 5 per cent to 100 per cent of incoming stream flow under thirty-three climate change scenarios. These studies provide stochastically robust evaluations of filling strategies while considering the potential effects of climate change during the filling period, however the water resource models used only adapt GERD operations and thus do not consider potential strategic coordination with the other reservoirs.

Although many of the previously published studies analyse a number of possible filling strategies and provide insight to their implications, there remains an urgent need for a robust analytical policy-oriented modelling framework that sufficiently represents all major reservoirs and water uses in the system. A model used for policy-making must adequately simulate the existing operational practices of the system and incrementally incorporate the many potential filling alternatives that may emerge through the process of negotiation. Equally as important, the model must be physically and cognitively accessible to be trusted by the parties participating in the negotiations. As described by Olsson and Andersson (2007), the ability of stakeholders to understand and criticise the methods and assumptions embedded in models determines the acceptance or rejection of the results they provide. This supports the argument by Cash *et al.* (2003), who claim the essential role of knowledge systems to enhance the credibility, legitimacy and saliency of the information they provide. In the context of models for supporting negotiations over contested rivers, we put forth the concept of a 'hydro-policy' modelling framework that should be robust enough to represent a system with sufficient *accuracy*, be sufficiently *flexible* in its architecture, and be *transparent* enough to be understood and trusted by stakeholders. Furthermore, any study should demonstrate a completeness of sampling hydrology and management strategies.

Method

Modelling framework

The RiverWare platform was selected to develop and test various potential filling strategies for the GERD due to its flexible programming architecture (Zagona *et al.* 2001). This modelling software employs an object-oriented user interface to mimic the principal physical components that affect water management in the basin. Engineering algorithms are selected from extensive libraries that compute facets of water management for each object based on the known physical characteristics of the system. Data resulting from user inputs and these algorithms are propagated throughout the model network through links, typically resulting in an under-determined system wherever management decisions must be made. The prioritised rule-based simulation then provides the model objects with additional scripted user input which characterises the myriad of multi-objective operational policies that govern the management of water including international and intra-national agreements between users, water rights arrangements, legal constraints and dam management guidelines. This capability has been demonstrated by its recent successful use in transboundary negotiations over international management of the Colorado River (United States of America and United Mexican States 2012).

Nile Basin model structure

The major features in the Eastern Nile Basin that significantly affect water management and distribution are included as objects in the modelling workspace. The calibration used the historical time period from 1900 until 2002 and included the following features: Lake Tana with the Tana-Beles Hydropower Project and Tekeze Reservoir in Ethiopia; Roseires, Sennar, Jebel Aulia, Khashm el Girba and Merowe reservoirs in Sudan; and Lake Nasser/Lake Nubia formed by the HAD in Egypt. Within the calibration simulation, each of the constructed features are invoked at the point in time when construction was completed. When simulating future conditions, the recently heightened Roseires Dam, the newly developed Upper Atbara and Setit Dam complex, and the GERD are included as well. Monthly naturalised hydrologic input locations include 162 inflow nodes within South Sudan, Ethiopia, Sudan and Egypt. Demand locations reflect the major Sudanese diversion structures of the Gezira-Managil, New Halfa and Rahad schemes as well as the minor diversions from the Jebel Aulia reservoir and small aggregated demands between gauged locations. Consumptive or non-consumptive water uses within Egypt are not modelled beyond expected monthly releases from Aswan and necessary spills into the Toshka diversion works.

Data requirements

Naturalised historical hydrologic inflow data – meaning non-depleted and unregulated by anthropogenic effects – for the 103 years (1900–2002) was developed by van der Krogt and Ogink (2013) and provided by the ENTRO office of the NBI. This study compiled data collected from a variety of sources with differing periods of record and filled in missing data using site-specific regression and partitioning techniques. The index-sequential method (Kendall and Dracup 1991; Ouarda et al. 1997) was used to develop stochastic hydrologic conditions by applying the historical sequence to the future modelled period (2016–2059) with a start date that corresponds to each of the years in the reconstructed historical record. The length of the simulation period was selected to allow the model to reach equilibrium after the effects of filling under all hydrologic conditions. We acknowledge that the selected hydrologic method does not reflect future transient climate change conditions (Milly et al. 2008), the *Hurst Effect* of persistent behaviour of flows (Hurst et al. 1965), or recent understanding of low-frequency oscillations (Zhang et al. 2016), however, the approach is considered sufficiently robust for this analysis, given the short-term nature of the filling process.

Current water use estimates were obtained from the first phase of the Nile Basin DSS development (Carron *et al.* 2011). Historical diversions were also obtained from the Nile Encyclopaedia (Nile Control Staff 1933–Present) and used for calibration when available. Diversions for the Gezira/Managil scheme were updated with recent monthly averages from September 2012 to August 2014 (Sudan MoWE 2015). To isolate the effects directly attributable to the GERD from increases to water use that could be indirectly attributable to the GERD, current diversion volumes were assumed to remain constant throughout the future modelled period. For the purposes of this study, Ethiopia's diversions from the Nile are assumed to remain insignificant, while Sudan's diversions are assumed to remain at the current estimated volume of 16.0 BCM per year using the data sources described above. Egypt's annual diversions are assumed to be 55.5 BCM, which is equivalent to the allocation specified in the 1959 Treaty with Sudan (Nile Treaty 1959).

Historical and current reservoir operations were acquired from numerous information sources throughout the basin including the Nile Basin Initiative (ENTRO 2013), water resource agencies (Egypt MWRI 2005; Sudan MoIHP 1968), engineering reports (Lahmeyer International 2005; PB Power 2003; Salini Costructtori 2006; SMEC International 2012) and analysis of various existing models (Jonker *et al.* 2012; Mulat and Moges 2014; van der Krogt and Ogink 2013; Yao and Georgakakos 2003). In the absence of further information, operations of single purpose hydropower reservoirs were simulated to generate a target power production. Reservoirs with more available information were simulated as being managed for a combination of hydropower production, control of sediment accumulation by seasonal reduction of pool elevations, satisfying irrigation diversions from the reservoir and meeting downstream flow requirements (Table 10.1). Prioritised operation rules were written to simulate the changing objectives of the reservoirs throughout the year. The model was made available and refined through seven training workshops conducted between 2012 and 2016 at NBI-ENTRO, Addis Ababa University, University of Khartoum, Cairo University, Sudan's Dams Implementation Unit of the Ministry of Water Resources, Irrigation and Electricity and Egypt's Water Resources Research Institute of the Ministry of Water Resources and Irrigation. After operational policies were simulated in the rules with the best available information, modifications were made to explore alternative management policies. All rules in the model were transparently written to allow the technically trained stakeholders to readily understand the operational logic. Participation in these training sessions did not imply endorsement of the model or the results.

Table 10.1 Prioritised operation parameters of existing reservoirs

Month	Roseires Min. outflow (MCM) [1]	Roseires Downstream demands (MCM) [2]	Roseires Target elevation (m) [3]	Roseires Power generation (MW) [4]	Sennar Min. outflow (MCM) [1]	Sennar Direct diversions (MCM) [2]	Sennar Target elevation (m) [3]	Sennar Power generation (MW) [4]	Jebel Aulia Direct withdraws (MCM) [1]	Jebel Aulia Target elevation (m) [2]	Jebel Aulia Power generation (MW) [3]	Khashm El Girba Min. outflow (MCM) [1]	Khashm El Girba Direct diversions (MCM) [2]	Khashm El Girba Target elevation (m) [3]	Khashm El Girba Power generation (MW) [4]	Merowe Min. outflow (MCM) [1]	Merowe Target elevation (m) [2]	Merowe Target power (MW) [3]	HAD Target release (MCM) [1]	HAD Power generation (MW) [2]
Jan		*	—			856	421.7		56	377.4			113	474			—	470	3,510	
Feb		*	—			794	421.7		60	377.4			96	474			—	536	3,920	
Mar		*	—			416	421.7		75	376.9			74	474			—	565	4,380	
Apr		*	—			69	421.7		81	375.4			44	474			—	639	4,120	
May		*	—			138	417.0		88	373.9			52	474			290	699	5,080	
Jun	← 1240 MCM/month →	—	470	← By-product of other operations →	← 372 MCM/month →	659	417.0	← By-product of other operations →	92	372.5	← By-product of other operations →	← 62 MCM/month →	106	464	← By-product of other operations →	← 1,875 MCM/month →	290	861	6,320	← By-product of other operation →
Jul		—	469			937	417.0		185	376.5			122	464			290	860	6,760	
Aug		—	469			474	417.0		193	376.5			138	464			290	740	5,900	
Sep		—	487			938	418.7		352	377.4			134	474			298	861	4,150	
Oct		—	490			1,040	421.7		418	377.4			148	474			300	860	3,890	
Nov		*	—			926	421.7		351	377.4			121	474			300	625	3,840	
Dec		*	—			922	421.7		99	377.4			121	474			300	511	3,720	
Priority	1	2	3	4	1	2	3	4	1	2	3	1	2	3	4	1	2	3	1	2

Notes

* Roseires release to meet storage proportional percentage of ((Gezira + DS Sennar Demands) + Roseires to Sennar Demands − Roseires to Sennar Inflows.

Tekeze Operated for 112 MW constant hydropower, Upper Atbara and Setit Dam complex operated for 43 MW constant hydropower.

Reservoir filling scenarios

General approach

The goals of this study were to identify and evaluate potential filling and management options, and to test the major dimensions of water distribution to, and energy production from, the three countries during the filling period. Drawing from the experience of previous studies (Bates *et al.* 2013; IPoE 2013; King and Block 2014; Zhang *et al.* 2015), many filling strategies were envisaged and tested. Two general paradigms of GERD management during filling emerged: (1) reach an agreement for the GERD to release a minimum flow or volume over time; and (2) adopt a specified or capped filling rate over time. Only the minimum flow paradigm allows the GERD to fill faster during wet years and slower during dry years while meeting a minimum water requirement for Sudan and Egypt (MIT 2014), so this paradigm is the focus of the results reported herein.

Common assumptions

Certain characteristics of the GERD filling were assumed for all scenarios based on stated criteria of the chief dam construction engineer, S. Bekele,[1] and known physical characteristics of the GERD (IPoE 2013; MIT 2014). The reservoir is assumed to fill during the initial year (2016) to 560 m (3.58 BCM) to test the first two installed turbines and remain at that elevation until the start of the second year flood period (2017). Additional turbines are assumed to come online every two to three months. Downstream releases may be passed through the increasing number of installed turbines, through bottom outlets or over the incrementally raised open spillway. Starting in 2017, monthly releases patterns from the GERD during the filling period are assumed to evenly distribute an agreed annual release volume throughout the year to the extent possible (Ethiopian NPoE 2013), while readjusting continuously if shortfalls are encountered. Once the minimum operation level of 590 m (14.7 BCM) is reached, maintaining this level is assumed to take priority over downstream releases. The filling is considered complete when the reservoir level reaches 640 m (74.0 BCM) (Mulat and Moges 2014) at which time a policy of regular energy generation of 1,308 GWh/month begins.[2] All of these assumptions were based on best available knowledge at the time of the analysis.

The HAD is assumed to be operated primarily to meet downstream demands that total 55.5 BCM per year. The minimum elevation for power generation and downstream releases is 147 m (31.9 BCM), and the elevation range from 175 to 182 m (121 to 167 BCM) is reserved for emergency storage or flood protection operations. Pool elevations above 178 m are assumed to begin spilling into the Toshka canal (van der Krogt and Ogink 2013). A drought management policy reduces deliveries to downstream water users by 5 per cent as the storage volume in Lake Nasser decreases below 60 BCM (159.4 m), 10 per cent when storage falls below 55 BCM (157.6 m), and 15 per cent when storage falls below 50 BCM (155.7 m) (Egypt MWRI 2005; also[3]).

Scenarios analysed

The model was used to study the effects of modifying five factors within the system. Some items are simple numerical changes that represent the sensitivity of a particular parameter, while others represent conceptual changes to operation policies that respond to the existence of the GERD. The five factors analysed in this paper are:

1   *Total agreed annual release volume from the GERD during the filling period* – Six agreed annual release values of 25 BCM, 30 BCM, 35 BCM, 40 BCM, 45 BCM and 50 BCM per year were analysed to reflect the range from below the 1984 drought flow (30.9 BCM) to above the average annual flow (49.4 BCM). In addition, a rapid fill scenario (0 BCM) was also analysed for comparative purposes. The GERD will attempt to release this volume every year until the filling is complete using the assumptions provided above.
2   *Starting conditions for the HAD* – A range of four starting pool elevations of 165 m, 170 m, 175 m and 180 m is used to demonstrate the possible effects in Egypt resulting from initial conditions.
3   *Sudan Reservoir Operations* – Two potential scenarios were simulated including: (a) all reservoirs use current operation rules; and (b) reservoirs are operated at the maximum elevation feasible with releases only to meet hydropower demands (Merowe), meet downstream demands (Roseires and Sennar) and flood control operations, thus forgoing seasonal flushing for sediment.
4   *HAD Drought Management Policy* – Two potential HAD operation scenarios were simulated including: (a) no drought management policy implemented; and (b) the drought management policy implemented that reduces downstream deliveries based on low storage thresholds as described above.
5   *GERD–HAD Safeguard Policy* – A policy was envisaged that uses the storage in the GERD to ensure that the minimum power pool elevation of the HAD (147 m) is protected. This alternative evaluates whether the pool elevation of the HAD is expected to fall below 150 m (providing a 3 m buffer), and if so, an additional release is made from the GERD to try to maintain this elevation. This additional release is made after any decision to implement the HAD drought management policy is made and thus reduced HAD releases are maintained. This policy terminates when the GERD reaches 640 m. Two potential scenarios were simulated including: (a) no GERD support of the HAD; and (b) with the GERD explicitly supporting the HAD with the above criteria. Additional thresholds and release volumes can be explored.

The five dimensions described above were used to generate 224 combinations of policies and initial conditions, each being subject to 103 hydrologic traces and thus requiring around 23,000 simulations.

Results

Time to fill the reservoir

A key metric across the potential scenarios is how much time would be required to fill the GERD. Figure 10.1 demonstrates the increase in average time to fill and the variance in that time given hydrologic variability with an increasing agreed annual release. Including the first year of fill to 560 m for testing the turbines, the fastest possible time to fill the reservoir to the full supply level of 640 m would be during the flood of the third year (2018). In situations where the agreed annual release exceeds the average annual flow rate of 49.4 BCM, the reservoir on average cannot fill completely.

Effects on downstream consumptive uses

A major concern of downstream countries is whether the GERD will negatively affect the reliability of their water supply. The current management practices of the Blue Nile reservoirs in Sudan are designed to operate the reservoirs at a minimum elevation to pass sediment until late September, and then capture the

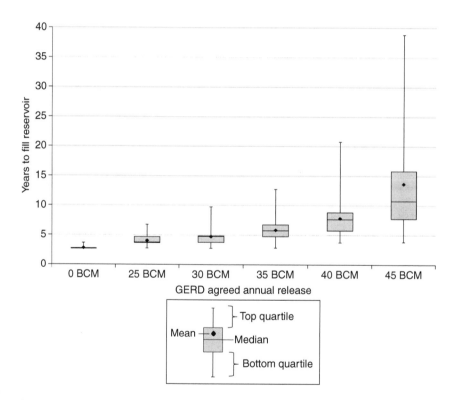

Figure 10.1 Years required to fill the GERD under various agreed annual GERD releases.

end of the flood flow to retain sufficient storage to meet the needs of the Gezira-Managil diversion. The results show that this current operation plan is not compatible with the assumed near constant releases of the GERD during filling. However, by starting the Sudanese reservoirs at the maximum capacity when the filling of the GERD begins and reoperating them to make releases only to meet downstream demands and allow necessary spills during flooding, the risk of shortages to Sudanese irrigated agriculture and municipal uses is essentially eliminated. This reoperation may be feasible due to sediment capture of the GERD, but warrants further investigation.

Shortages to Egyptian water users are highly subject to the hydrologic conditions during the filling period, the initial storage volume behind the HAD, as well as the agreement reached between the countries. As an example of a relatively bleak scenario, if no agreement is reached with Ethiopia to release water from the GERD during filling and the pool elevation of the HAD is relatively low at 165 m when the filling begins, shortages to Egyptian water users were shown to occur in approximately 40 per cent of the hydrologic scenarios by the third year of filling. Approximately 40 per cent of these shortages exceeded 10 BCM and 10 per cent exceeded 20 BCM. While this unilateral filling policy under low HAD initial storage conditions may not result in any shortages to Egypt under average hydrologic conditions, this approach demonstrates that substantial risks to Egypt do exist, if dry conditions occur and agreements are not reached. This level of risk clearly highlights the need for the countries to coordinate.

By assuming 175 m as the starting pool elevation of the HAD at the start of the filling period, Figure 10.2 demonstrates the probability of exceedance of shortages to Egypt during the initial ten years after filling commences and across the various agreed annual releases for all four combinations of inclusion and exclusion of both the HAD drought management policy and the GERD–HAD safeguard policy. While these cumulative plots demonstrate the potential range of shortages across policies and their relative probabilistic distributions over the time period, they do not reflect specific annual risks as described in the scenario above. Figure 10.2a shows that significant shortages are possible under dry conditions, if neither the HAD drought management policy or GERD–HAD safeguard policy are put in place. Figure 10.2b demonstrates the effect of implementing the HAD drought management policy which reduces the risk of severe shortages while increasing the likelihood of proactive reductions to water users, but also demonstrates that the risk of shortages beyond these planned reductions are not eliminated for agreed annual releases of 30 BCM and less. Figure 10.2c shows that the GERD–HAD safeguard policy by itself does mitigate the vast majority of the risk without the need for proactive reduced deliveries, however, the possibility of high magnitude shortages remain under extreme conditions due to the minimum operation level of 590 m for the GERD and the assumed immediate termination of the policy after filling. In Figure 10.2d, the combination of both the HAD drought management policy and the GERD–HAD safeguard policy is shown to completely eliminate the risk of

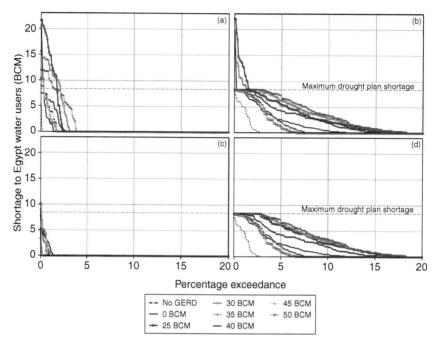

Figure 10.2 Cumulative probability of exceedance of annual shortages to Egypt across 2016–2025 with initial pool elevation of HAD = 175 m. (a) No HAD Drought Policy, no GERD-HAD Safeguard Policy; (b) with HAD Drought Policy, no GERD-HAD Safeguard Policy; (c) no HAD Drought Policy, with GERD-HAD Safeguard Policy; (d) with HAD Drought Policy, with GERD–HAD Safeguard Policy.

Note
GERD = Grand Ethiopian Renaissance Dam; HAD = High Aswan Dam.

unplanned shortages to Egyptian water users. Each of these plots demonstrates the paradox that higher agreed annual releases provide a guaranteed delivery downstream during filling, but also prolong the filling process and therefore extend the risk to downstream users.

Table 10.2 highlights the differences between the policies in more critical terms by calculating the maximum probability of the HAD reaching 147 m across all points in time throughout the forty-three-year model run period. Below this elevation power cannot be produced and requested downstream releases cannot be made therefore the model reduces downstream releases to maintain this elevation. Table 10.2a shows this probability, if neither the HAD drought management policy nor the GERD–HAD safeguard policy are used. Table 10.2b shows the degree to which the assumed HAD drought management policy cannot alone protect the minimum pool elevation, and Table 10.2c shows the extent to which the GERD–HAD safeguard policy alone leaves a

Table 10.2 Maximum probability of High Aswan Dam reaching the minimum power production elevation (147 m) under four management scenarios across the run period (2016–2059)

		(a) No HAD Drought Mgmt. Policy, No GERD-HAD Safeguard Policy				(b) With HAD Drought Mgmt. Policy, No GERD-HAD Safeguard Policy			
		Initial HAD elevations				*Initial HAD elevations*			
		180 m	175 m	170 m	165 m	180 m	175 m	170 m	165 m
No GERD		0%	0%	0%	0%	0%	0%	0%	0%
Agreed Annual Release	50 BCM	3%	4%	6%	10%	0%	0%	0%	2%
	45 BCM	3%	4%	6%	11%	0%	0%	0%	3%
	40 BCM	5%	6%	7%	13%	0%	0%	0%	5%
	35 BCM	7%	9%	15%	31%	1%	1%	3%	6%
	30 BCM	7%	9%	21%	45%	4%	5%	6%	11%
	25 BCM	8%	9%	27%	47%	6%	7%	9%	18%
	0 BCM	2%	7%	17%	37%	1%	3%	10%	20%

		(c) No HAD Drought Mgmt. Policy, With GERD-HAD Safeguard Policy				(d) With HAD Drought Mgmt. Policy, With GERD-HAD Safeguard Policy			
		Initial HAD elevations				*Initial HAD elevations*			
		180 m	175 m	170 m	165 m	180 m	175 m	170 m	165 m
GERD Agreed Annual Release	50 BCM	2%	4%	5%	8%	0%	0%	0%	1%
	45 BCM	2%	3%	4%	8%	0%	0%	0%	1%
	40 BCM	2%	3%	3%	7%	0%	0%	0%	2%
	35 BCM	2%	2%	3%	7%	0%	0%	0%	1%
	30 BCM	2%	2%	3%	5%	0%	0%	0%	2%
	25 BCM	2%	3%	4%	7%	0%	0%	1%	1%
	0 BCM	2%	3%	5%	21%	0%	0%	2%	2%

remaining risk due to limitations of the minimum operating level of the GERD and immediate termination of the policy. When applying either of these policies independently, it can be seen that much of the risk of reaching this critical elevation is reduced, but some risk still remains. Finally Table 10.2d shows that the combination of the HAD drought management policy and the GERD–HAD safeguard policy does provide almost complete protection for the HAD.

While a unilateral approach to reservoir operations poses a risk to downstream countries, it is also likely that the countries will reach some level of agreement that assures a minimum release from the GERD in each year. Across all starting conditions of the HAD, the results indicate that agreed annual releases from the GERD of 35 BCM and greater significantly reduces the average shortage volume, while releases of 30 BCM or less may be insufficient to keep up with the 55.5 BCM annual Egyptian demand. Perhaps even more revealing are the benefits of risk reduction through close and continuous coordination throughout the filling period. Adapting releases from the GERD if the risks of downstream shortages increase is one tangible example of continuous coordination. As the effects of filling subside, the long-term risk of shortages for Egypt decreases relative to the baseline condition due to the benefit of flow regulation that the GERD provides. However, future water resource developments within Sudan are likely to become a greater concern for water supply to Egypt over the long term.

Effects on hydropower generation

Table 10.3 reports the average annual change to Ethiopian, Sudanese and Egyptian hydropower generation with the addition of the GERD averaged across two time frames: the initial ten years starting from the commencement of filling (short term) and the following twenty years (medium term). The results for Sudan assume adaptations to Sudanese reservoir management takes place, and the results for Egypt assume both the HAD drought management policy and the GERD–HAD safeguard policy.

The average annual energy generation in Ethiopia (dominated by the GERD) increases with decreased agreed annual release in the short term, which is largely based on completing the filling early and transitioning to normal operations. Sudanese energy generation is improved, particularly when strategically reoperating their reservoirs. In contrast, the combination of the HAD drought management policy and GERD–HAD safeguard policy results in competing factors that result in increases and decreases to energy generation for Egypt. These include reduced HAD turbine flows, additional water made available from the GERD, increased pool elevation, and the reduced likelihood of reaching the minimum power generation elevation.

Various non-linear behaviours can be observed in Table 10.3 that are largely based on the timing for which filling is achieved relative to the time periods used for averaging. The maximum energy generation in Ethiopia occurs at the point of transition to normal operations when the GERD is at 640 m; therefore,

Table 10.3 Change of average annual energy generation (GWh/year) due to the GERD with downstream adaptations

		Ethiopia		Sudan: with reops		Egypt: with drought management; with GERD-HAD safeguard							
		Short-term effect	Medium-term effect	Short-term effect	Medium-term effect	Short-term effect				Medium-term effect			
						Initial HAD pool elevation				Initial HAD pool elevation			
						180m	175m	170m	165m	180m	175m	170m	165m
GERD Agreed Annual Release	50 BCM	10,339	13,481	1,498	2,262	−869	−943	−1,024	−1,030	−342	−369	−425	−456
	45 BCM	10,660	14,011	1,360	2,234	−1,086	−1,167	−1,259	−1,259	−309	−341	−399	−425
	40 BCM	11,106	14,037	1,216	2,244	−1,309	−1,393	−1,469	−1,443	−211	−233	−267	−275
	35 BCM	11,441	13,824	1,158	2,246	−1,384	−1,459	−1,507	−1,460	−139	−158	−188	−194
	30 BCM	11,662	13,556	1,081	2,227	−1,370	−1,440	−1,482	−1,430	−121	−138	−160	−169
	25 BCM	11,780	13,477	996	2,223	−1,356	−1,426	−1,456	−1,403	−118	−131	−155	−162
	0 BCM	11,891	13,306	952	2,206	−1,124	−1,198	−1,233	−1,181	−98	−116	−137	−141

Note
* Short term = average from 2016–2025 (10-years), medium term = average from 2026–2045 (11–30 years).

this peak occurs in the short term under low agreed annual releases and the medium-term period for higher releases. Similar behaviour of Sudanese and Egyptian power generation can be seen as well. One notable result is the non-linearity from the 50 BCM agreed annual release scenario, which is due to this release exceeding the average annual flow of 49.4 BCM resulting in the GERD not being able to reach full capacity on average and consequentially less medium-term energy generation for Ethiopia and slightly greater generation for Egypt and Sudan.

Table 10.4 provides another perspective to demonstrate the foregone energy benefit to Ethiopia to provide assurance to Egypt through the GERD–HAD safeguard policy. The exceedance probability of generating the target energy of 1,308 GWh/month is shown as a function of the agreed annual delivery and starting elevation of the HAD. This table again demonstrates that the ability to meet this power generation target is essentially unchanged when protecting the HAD power pool elevation given any additional release from the GERD can be passed through the turbines to generate electricity. Assumptions were made regarding the ability of the GERD to generate energy throughout the filling period, while making the necessary releases and the uniform energy demand pattern after the filling is complete, however, the demand for energy will depend on factors such as the local market demand, capacity of transmission lines, and timely completion and/or expansion of the regional interconnection projects (Block and Strzepek 2010; MIT 2014). This highlights the opportunity to match the agreed annual release with these energy demands and the need for flexible energy demands if additional releases are required.

Discussion

The results of this study confirm that a substantial amount of hydropower will be produced by the GERD once the turbines are installed and the reservoir begins to fill. In comparison, lesser positive and negative effects to energy production in Sudan and Egypt respectively may occur during this period. Assuming future hydrologic conditions are similar to historical sequences and the current rates of water consumption continue during the filling period, the probability of shortages to existing downstream consumptive uses will temporarily increase during the filling period of the GERD in the absence of an agreement between the countries. However, these risks of shortages can be minimised, if not eliminated, if the countries are able to reach an agreement on how the filling will occur. The combination of an agreed annual release from the GERD, proactive reoperation of the Sudanese reservoirs, implementation of a drought management policy for the HAD, and a safeguard release from the GERD if the HAD pool elevation falls below a critical level offers one possible solution to manage these risks. Assured protection of Egypt's needs across all hydrologic conditions is only feasible with cooperative management of the upstream infrastructure in Ethiopia and Sudan. Continuous sharing of information between the countries is essential to the implementation of any coordinated management solution.

Table 10.4 Change in reliability of a 1,308 GWh/month firm energy generation of the GERD due to implementation of the GERD-HAD Safeguard Policy

| | | Short term: 2016–2025 | | | | Medium term: 2026–2045 | | | | |
| | | No safeguard Initial HAD elevations | | | | No safeguard Initial HAD elevations | | | | |
		180 m	175 m	170 m	165 m		180 m	175 m	170 m	165 m
GERD Agreed Annual Release	50 BCM	0.0%	0.0%	0.0%	0.0%	34.1%	0.0%	0.0%	0.0%	0.0%
	45 BCM	0.0%	0.0%	0.0%	0.1%	65.6%	0.0%	0.0%	0.0%	0.0%
	40 BCM	0.0%	0.0%	0.0%	0.2%	79.5%	0.0%	0.0%	0.0%	0.0%
	35 BCM	0.0%	0.0%	-0.1%	0.1%	80.0%	-0.1%	-0.1%	-0.1%	-0.1%
	30 BCM	0.0%	0.0%	-0.5%	-0.7%	78.2%	0.1%	0.1%	0.2%	0.2%
	25 BCM	-0.2%	-0.4%	-0.6%	-1.2%	77.6%	0.1%	0.1%	0.2%	0.3%
	0 BCM	0.0%	-0.1%	-0.3%	-0.8%	76.5%	0.0%	0.0%	0.0%	0.0%

Modifications to the operations of the Roseires and Sennar dams to accommodate the changes to the intra-annual timing of releases from the GERD will be necessary to provide reliable water supplies to the large irrigated agricultural areas of Sudan. This will require the Roseires and Sennar reservoirs to reach their full capacity during the first year of GERD filling and make releases to only meet direct diversion requirements to the Gezira/Managil canals, satisfy the minimum downstream flow requirements, and pass any flood water while retaining the maximum possible volume of storage. However, without agreed annual GERD releases and proper re-operation of Sudanese reservoirs, significant losses to energy generation in Sudan may occur in the initial years. Once filling is complete, up to 21 per cent increases in energy generation can result due to greater available flows during the non-flood period and reduction of spills during the flooding season. Not only will these planning changes be necessary to assure water security, but close coordination between the Blue Nile reservoirs will be critical to ensure safety from upstream releases overtopping downstream dams. Coordination must include effective communication between the operators of the GERD, Roseires and Sennar dams to plan their releases accordingly. Implications of sediment management during this transitional period were not analysed in this study and we did not attempt to evaluate the impacts to flood recession agriculture in Sudan, either along the river or around the reservoirs due to revised dam operations.

Several factors affect the potential risks to Egyptian water supplies and energy production during the filling phase of the GERD. These include the initial storage in Lake Nasser when filling begins, the hydrologic conditions that occur during the filling period, an agreed annual release from the GERD, and the operational policies of the HAD and all upstream reservoirs. This study quantifies this risk across each of these dimensions and demonstrates that the potential for impacts to Egypt be significantly reduced with proper planning and coordination. The HAD provides a buffer of water to meet Egypt's needs, however, if the pool elevation falls below the intake elevation of 147 m, Egypt will simultaneously lose both hydropower generation from the HAD and the ability to fully satisfy downstream needs. In this study, management of this risk was analysed across different agreed annual releases from the GERD and by examining two policies including: (1) Egypt proactively reducing releases through the HAD current drought management policy; and (2) Ethiopia making additional releases when the elevation of the HAD is expected to fall below a pre-specified trigger elevation and Sudan allowing this water to pass downstream to Egypt.

By relying only on an agreed annual release, the risk of reaching this minimum elevation ranges from 2 per cent to 47 per cent, depending on the release value and the initial storage of the HAD (Table 10.2a). Implementing the HAD drought management policy reduces the risk of abrupt and potential large shortages by making carefully planned reductions to releases and downstream diversions (Figure 10.2b and Table 10.2b). While Egypt may argue that invoking their drought management policy to offset the impacts of the GERD is not acceptable, the willingness to do so provides Ethiopia with the sound rationale to provide further protection to Egypt if a drought occurs during the filling period. A notable degree of risk remains, even with the HAD drought policy in place.

In contrast, the GERD–HAD safeguard policy endeavours to maintain the HAD at an elevation of 150 m regardless of an agreed annual release. In this case, the extra volume released is dynamically estimated to assure the 150 m pool elevation is maintained based on the expected incoming hydrology and downstream Egyptian demands. However, due to losses, lags, extreme hydrologic conditions and infrastructure limitations, maintaining this level is not always certain. The results indicate that the GERD–HAD safeguard policy alone largely protects the HAD and avoids the need for Egypt to proactively reduce releases downstream of the HAD. However, the minimum power elevation of the GERD can be a limiting factor when providing this supplemental water and Egypt's risk persists after the GERD reaches maximum elevation. This policy may need to be extended for a period of time after the GERD is filled to assure the risk to Egypt is alleviated.

To eliminate effectively all risks of the HAD reaching the minimum power elevation, a combination of the HAD drought management policy and the GERD–HAD safeguard policy was shown to be effective (Figure 10.2d; Table 10.2d). While this policy suggests potential proactive reductions to Egypt's deliveries from the HAD, it avoids the risk of unplanned shortages to Egyptian water users. Support from the GERD can be made to maintain the pool elevation of 150 m after assuming planned downstream releases subject to the HAD drought management policy. Any such collaboration between the GERD and the HAD requires an increased level of cooperation that assures particular releases have been made from each reservoir. This highlights the need for continuous information sharing, which requires agreed measurement procedures, establishes protocols to assure data quality and plans for appropriate dispute resolution mechanisms.

For the GERD to provide safeguard releases to the HAD, a small foregone generation energy from the GERD may occur relative to the generation capacity as a whole by making releases that do not pass through the turbines. However, this largely depends on Ethiopia's ability to utilise the energy generated whenever the water is needed downstream. If there is demand and transmission capacity that can absorb the energy generated when these excess flows are required, then protecting the HAD with flows from the GERD can be economically beneficial (Tawfik 2016). In addition to the use of the HAD drought management policy, the three key additional components of this strategy: an agreed annual release, a trigger elevation for protecting the critical HAD power pool elevation, and the calculation of a safeguard release volume, are subjects for negotiation and further analysis.

Conclusions

A substantial amount of dialogue has taken place regarding the GERD and several previous analyses have attempted to quantify the potential benefits and impacts to Ethiopia, Sudan and Egypt. However, there remains an urgent need for the countries to reach an agreement on specific strategies of how to manage the process of filling the reservoir. We demonstrated that many potential approaches exist which must be examined under a variety of potential hydrologic scenarios to understand their probabilistic benefits and risks. Coordination is a continuum ranging from unilateral decisions to dynamic operations that continuously reflect current conditions and an awareness of the water security situation between co-riparians. The risks to the downstream countries of Sudan and Egypt can decrease significantly if the countries are willing to work closely together through proper planning and information sharing. In this study, we present some possible arrangements of reservoir coordination to achieve this goal, and we describe an analytical framework that quantifies the benefits and risks. An agreed annual release from the GERD can reduce downstream risks, however, further along this continuum lies a joint willingness to fill the reservoir in an efficient way that effectively assures water security downstream during filling. This study demonstrates this concept by analysing safeguard releases from the GERD to support the HAD under critical circumstances which seeks to eliminate the severe risks and impacts.

Risks to water supply and hydropower generation have always existed on the Nile and changes to the system may alter this risk profile either positively or negatively – and either temporarily or permanently. This study, along with others that consider the filling of the GERD, can provide important technical information for the negotiation process. A single correct solution is unlikely from any study, but the analysis allows negotiators to understand how significant the changes of risks might be, whether they are acceptable, how they might be managed and whether alternative approaches must be pursued. Ultimately, we believe this study demonstrates that a middle ground does indeed exist.

Notes

1 From personal communication with S. Bekele by author, 12 June 2015.
2 In the absence of future estimated energy demands patterns, 1,308 GWh/month represents the projected 15,692 GWh/year energy generation distributed evenly over the year (IPoE 2013). While the reservoir will have a 6,000 MW installed capacity and hence the ability to provide peak power generation and avoid all spills, the assumption used will only require power generation to exceed 2,000 MW in less than 2 per cent of all cases.
3 From personal communication with K. Hamed by author, 2012.

References

Arjoon, D., Mohamed, Y., Goor, Q. and Tilmant, A. (2014). Hydro-Economic Risk Assessment in the Eastern Nile River Basin. *Water Resources and Economics*, 8, 16–31. Doi: 10.1016/j.wre.2014.10.004.

Bates, A., Tuncok, K., Barbour, T. and Klimpt, J.-É. (2013). *First Joint Multipurpose Program Identification: Strategic Perspectives and Options Assessment on the Blue Nile Multipurpose Development – Working Paper 2*. Addis Ababa: Author Report to Nile Basin Initiative.

Block, P.J. and Strzepek, K. (2010). Economic Analysis of Large-Scale Upstream River Basin Development on the Blue Nile in Ethiopia Considering Transient Conditions, Climate Variability, and Climate Change. *Journal of Water Resources Planning and Management*, *136*(2), 156–166. Doi: 10.1061/(ASCE)WR.1943-5452.0000022.

Carron, J., Parkin, G., O'Donnell, G.M. and O'Connell, P.E. (2011). *NBI Water Resources Planning and Management Project, Nile Basin Decision Support System (DSS): Data Processing and Quality Assurance, Pilot Application of the Nile Basin Decision Support System: Stage 1 – Development of Pilot Case Models*. Boulder, CO: Author Report to Nile Basin Initiative.

Cash, D.W., Clark, W.C., Alcock, F., Dickson, N.M., Eckley, N., Guston, D.H., Jager, J. and Mitchell, R.B. (2003). Knowledge Systems for Sustainable Development. *Proceedings of the National Academy of Sciences*, *100*(14), 8086–8091. Doi: 10.1073/pnas.1231332100.

DoP (2015). Agreement on Declaration of Principles between the Arab Republic of Egypt, The Federal Democratic Republic of Ethiopia and the Republic of the Sudan on the Grand Ethiopian Renaissance Dam Project (GERDP). Khartoum: State Information Service.

EDF and Scott Wilson (2007). *Eastern Nile Power Trade Program Study: Pre-Feasibility Study of Border Hydropower Project, Ethiopia*. Addis Ababa: Author report to Nile Basin Initiative.

Egypt MWRI (2005). *Water for the Future: National Water Resources Plan for Egypt – 2017*. Cairo: Ministry of Water Resources and Irrigation.

ENTRO (2013). Eastern Nile Planning Model – Analytical and Modeling Tools [CD-ROM] *Nile Cooperation for Results (NCORE) Workshop*. Eastern Nile Technical Regional Office, Addis Ababa: Nile Basin Initiative.

Ethiopian NPoE (2013). *Unwarranted Anxiety: The Grand Ethiopian Renaissance Dam (GERD) and Some Egyptian Experts Hyperbole*. Retrieved 4 April 2016, from www.mowr.gov.et/index.php?pagenum=0.1&ContentID=88.

Goor, Q., Halleux, C., Mohamed, Y. and Tilmant, A. (2010). Optimal Operation of a Multipurpose Multireservoir System in the Eastern Nile River Basin. *Hydrology and Earth System Sciences*, *14*(10), 1895–1908. Doi: 10.5194/hess-14-1895-2010.

Guariso, G. and Whittington, D. (1987). Implications of Ethiopian Water Development for Egypt and Sudan. *International Journal of Water Resources Development*, *3*(2), 105–114. Doi: 10.1080/07900628708722338.

Hassan, A. (2012). *Eastern Nile Planning Model (ENPM) Project: Water Balance Model for the Eastern Nile Basin*. Dhaka: Author report to Nile Basin Initiative.

Hurst, H.E., Black, R.P. and Simaika, Y.M. (1965). *Long-Term Storage: An Experimental Study*. London: Constable.

IPoE (2013). *International Panel of Experts on the Grand Ethiopian Renaissance Dam Project Final Report*. Addis Ababa: Author.

Islam, S. and Susskind, L. (2012). *Water Diplomacy: A Negotiated Approach to Managing Complex Water Networks*. New York: Routledge.

Jeuland, M. (2010). Economic Implications of Climate Change for Infrastructure Planning in Transboundary Water Systems: An Example from the Blue Nile. *Water Resources Research*, *46*(11), W11556. Doi: 10.1029/2010WR009428.

Jeuland, M. and Whittington, D. (2014). Water Resources Planning Under Climate Change: Assessing the Robustness of Real Options for the Blue Nile. *Water Resources Research*, *50*(3), 2086–2107. Doi: 10.1002/2013WR013705.

Jonker, V., Beuster, H., Sparks, A., Dobinson, L., Mullins, D., Palmer, R., … Tuncok, K. (2012). *Data Compilation and Pilot Application of the Nile Basin Decision Support System (NB-DSS): Scenario Analysis Report: Work Package 2: Stage 2 Scenario Analysis Report:*

Integrated Nile Basin (Report No. 7327/107486). Cape Town: Author Report to Nile Basin Initiative.

Keith, B., Ford, D.N. and Horton, R. (2017). Considerations in Managing the Fill Rate of the Grand Ethiopian Renaissance Dam Reservoir using a System Dynamics Approach. *The Journal of Defense Modeling and Simulation, 14*(1), 33–43. Doi: 10.1177/1548512916680780.

Kendall, D.R. and Dracup, J.A. (1991). A Comparison of Index-Sequential and AR(1) Generated Hydrologic Sequences. *Journal of Hydrology, 122*(1–4), 335–352. Doi: http://dx.doi.org/10.1016/0022-1694(91)90187-M.

King, A. and Block, P. (2014). An Assessment of Reservoir Filling Policies for the Grand Ethiopian Renaissance Dam. *Journal of Water and Climate Change, 5*(2), 233–243. Doi: 10.2166/wcc.2014.043.

Lahmeyer International (2005). *Feasible Study for the Merowe Irrigation Project.* Khartoum: Author report to Republic of Sudan Dams Implementation Unit.

McCartney, M.P. and Girma, M.M. (2012). Evaluating the Downstream Implications of Planned Water Resource Development in the Ethiopian Portion of the Blue Nile River. *Water International, 37*(4), 362–379. Doi: 10.1080/02508060.2012.706384.

Milly, P.C.D., Betancourt, J., Falkenmark, M., Hirsch, R.M., Kundzewicz, Z.W., Lettenmaier, D.P. and Stouffer, R.J. (2008). Stationarity is Dead: Whither Water Management?. *Science, 319*(5863), 573–574. Doi: 10.1126/science.1151915.

MIT (2014). *The Grand Ethiopian Renaissance Dam: An Opportunity for Collaboration and Shared Benefits in the Eastern Nile Basin: An Amicus Brief to the Riparian Nations of Ethiopia, Sudan and Egypt From the International, Non-Partisan Eastern Nile Working Group.* Boston, MA: Massachusetts Institute of Technology.

Mulat, A.G. and Moges, S.A. (2014). Assessment of the Impact of the Grand Ethiopian Renaissance Dam on the Performance of the High Aswan Dam. *Journal of Water Resource and Protection, 6*(6), 583–598. Doi: 10.4236/jwarp. 2014.66057.

NBI (2014). *Nile Basin Decision Support System DSS: Modeling Tools Training Module.* Entebbe: Nile Basin Initiative.

Nile Control Staff (1933–Present). *Nile Encyclopedia* [Data set]. Retrieved August 2014 from Nile Basin Initiative.

Nile Treaty (1959). *Agreement between the Arab Republic of Egypt and the Republic of the Sudan for the Full Utilization of the Nile Waters.* United Nations Treaty Series 64 1963.

Olsson, J.A. and Andersson, L. (2007). Possibilities and Problems with the use of Models as a Communication Tool in Water Resource Management. *Water Resources Management, 21*(1), 97–110. Doi: 10.1007/s11269-006-9043-1.

Ouarda, T.B.M.J., Labadie, J.W. and Fontane, D.G. (1997). Indexed Sequential Hydrologic Modeling for Hydropower Capacity Estimation. *Journal of the American Water Resources Association, 33*(6), 1337–1349. Doi: 10.1111/j.1752-1688.1997.tb03557.x.

PB Power (2003). *Building of Electricity Sector Database and Long-Term Power System Planning Study, Interim Report 3.* Newcastle: Author report to National Electricity Corporation of Sudan.

Sadoff, C.W. and Grey, D. (2005). Cooperation on International Rivers: A Continuum for Securing and Sharing Benefits. *Water International, 30*(4), 420–427. Doi: 10.1080/02508060508691886.

Salini Costructtori (2006). *Beles Multipurpose Project Level 1 Design General Report.* Rome: Author report to Ethiopian Electric Power Corporation.

SMEC International (2012). *Rosaries Dam Heightening Project (RDHP) Reservoir Operation Study – Revision 1 – Draft.* Dar es Salaam: Author report to Republic of Sudan Dams Implementation Unit.

Sudan MoIHP (1968). *Regulation Rules for the Working of the Reservoirs at Rosaries and Sennar*. Khartoum: Ministry of Irrigation and Hydro-Electric Power.

Sudan MoWE (2015). *Gaging Stations of the Blue Nile, White Nile, Main Nile, Upper Atbara, South of Sudan, Rahad and Dinder* [Data set]. Retrieved 3 September 2014 from Ministry of Water, Irritation and Electricity and available in author's personal file.

Tawfik, R. (2016). The Grand Ethiopian Renaissance Dam: A Benefit-Sharing Project in the Eastern Nile? *Water International*, 41(4), 574–592.

United States of America and United Mexican States (2012). Minute 319 – Interim International Cooperative Measures in the Colorado River Basin through 2017 and Extension of Minute 318 Cooperative Measures to Address the Continued Effects of the April 2010 Earthquake in the Mexicali Valley, Baja California. Washington DC: Government Printing Office.

van der Krogt, W. and Ogink, H. (2013). *Development of the Eastern Nile Water Simulation Model* (Report No. 1206020–000-VEB-0010). Delft: Author report to Nile Basin Initiative.

Wheeler, K. and Setzer, S. (2012). *Eastern Nile RiverWare Planning Model*. Addis Ababa: Author report to Nile Basin Initiative.

Yao, H. and Georgakakos, A.P. (2003). *The Nile Decision Support Tool, River Simulation and Management* (Report No. GCP/INT/752/ITA). Atlanta: Georgia Institute of Technology.

Yates, D., Sieber, J., Purkey, D. and Huber-Lee, A. (2005). WEAP21 – A Demand-, Priority-, and Preference-Driven Water Planning Model. *Water International*, 30(4), 487–500. Doi: 10.1080/02508060508691893.

Zagona, E.A., Fulp, T.J., Shane, R., Magee, T. and Goranflo, H.M. (2001). RiverWare: A Generalized Tool for Complex Reservoir Systems Modeling. *Journal of the American Water Resources Association*, 37(4), 913. Doi: 10.1111/j.1752-1688.2001.tb05522.x.

Zhang, Y., Block, P., Hammond, M. and King, A. (2015). Ethiopia's Grand Renaissance Dam: Implications for Downstream Riparian Countries. *Journal of Water Resources Planning and Management*, 141(9). Doi: 10.1061/(ASCE)WR.1943-5452.0000520.

Zhang, Y., Erkyihun, S.T. and Block, P. (2016). Filling the GERD: Evaluating Hydroclimatic Variability and Impoundment Strategies for Blue Nile Riparian Countries. *Water International*, 41(4), 593–610. Doi: 10.1080/02508060.2016.1178467.

Index

Addis Ababa 19, 25, 49–50, 55, 99, 103, 121–3, 129
Africa 46, 122–3, 129, 131, 181
Agreement on Declaration of Principles 31–2, 41, 51, 54, 56
agreements 4–6, 19–29, 31–3, 35–6, 41–4, 47–55, 68, 70–2, 95, 104–5, 107–8, 121–6; benefit-sharing 116; bilateral 92, 107, 120; comprehensive 95; cooperative 77; drafted by a transitional committee 72; historic 54; historical 125; intra-national 197; legal 78, 92, 104; military 122; multilateral 95; political 108; post-colonial 22; watercourse 64–5, 81
'Albert Nile' 18, 44
Anglo-Egyptian Sudan 19
Anglo-Ethiopian Nile Treaty 1902 5
annual flow 69, 186, 201, 208; rate of 202; total 20
Arab Republic of Egypt see Egypt
arrangements 42, 78, 115, 128, 131; benefit-sharing 79, 114, 120; bilateral 69; cooperative 73–4, 77, 79, 81–3; institutional 4; legal 64, 81; multilateral 107–8; treaty 64, 68, 81; water rights 197
Arsano, Y. 92, 99–100, 113, 117, 123
Artelia (French consultancy firm) 5, 54–5
Asegdew, G.M. 167, 169
Atbara River 18–19, 31, 43–4

balances 132, 143, 148–9, 162; asymmetric 102; between equitable and reasonable utilisation 73; of the Ethiopian–Sudanese rapprochement and convergence of positions on the GERD 122; general water 185–6; regional power 100

benefit-sharing 8, 12, 80, 113–14, 116, 120, 124, 126–7; and agreement on dam projects 116, 130–1; outcomes 12; perspective 116; as recommended by scholars 123, 132
benefits 4–5, 46–7, 56, 79–81, 83, 102–4, 113, 115–21, 124, 126–7, 131–2, 194; annual 139, 187; of cooperation and costs 114, 116, 119, 131–2; for the Eastern Nile 114; for Egypt 31, 161; of infrastructure development upstream 102; for Sudan 106; tangible 90, 93; in transboundary rivers 124
Bentham, Jeremy 66
Block, Paul 5, 9, 34, 52, 99, 139, 159–60, 181–90, 194–5, 200, 208
Blue Nile Basin 2, 50, 90, 94, 99, 129, 140, 158, 160, 174, 182; strategic 102; upper 181–2
Blue Nile River 1–2, 18–19, 31, 42–8, 54, 68, 77–9, 93–5, 98–9, 116–18, 138–40, 158–60; see also Blue Nile Basin
Boehlert, Brent 5, 8–10, 12–14, 79–81, 138, 138–56
Border dam 45, 56, 99, 117
borders 68, 97, 99, 121, 127, 140, 158, 160, 193
Britain 19–21, 41, 44; and Egypt 41, 44; and Ethiopia 41
BRLi Group 5
Brouwer, Roy 5, 9, 158–77
Brunnée, J. 25, 65, 70, 72, 82, 92, 95
Burundi 1–2, 18–19, 25–6, 29–30, 34–6, 41, 44, 52, 72–4, 95, 97–8, 122–3

Cairo 20, 26, 50, 94, 97, 104, 130
Calzadilla, A. 159, 162–4
Cascão, Ana Elisa 1–15, 53, 75, 90–108, 117, 119–20, 129, 131

Case Concerning Pulp Mills on the River Uruguay 66–7
CFA 6–7, 11, 13–14, 19, 25–31, 33–7, 72–9, 81–2, 91–2, 95, 97, 101–2; facilitators 27; negotiations 28, 30; processes 35, 101; signatures to 96, 101; subordinates 28; *see also* Cooperative Framework Agreement
CGE 8–9, 138–9, 146–7, 159; analyses of water infrastructure 161; methodology 159; models 138–9, 141, 143, 145–7, 149–50, 159, 161–2; multi-sector models 162; *see also* computable general equilibrium
Charter of the United Nations 71–2, 76
claims 4, 20–1, 23–4, 28–9, 35–6, 41–2, 47, 69, 107, 122, 196; by Egypt 20–1, 27, 29, 33, 41, 57; by Ethiopia 31; legal 65; non-negotiable 65; by the Sudan 21, 26–8; unqualified 14
climate change 4, 8–9, 15, 57, 61, 120, 177, 194, 196
Collins, R. 18–19, 41, 43
colonial treaties 19, 26, 28, 33, 69
'compliance pull' 62, 67, 70, 72, 74, 78–9, 81
Comprehensive Peace Agreement 97
compromise 14, 33, 36, 48–9, 53, 107, 125–7, 129, 132, 194; equitable 14; major 31; negotiated 12; solutions 10, 190; trade-offs and benefits of 194
computable general equilibrium 8–9, 138–9, 146–7, 159
concept of fairness 61–2, 66
cooperation 3–5, 7–15, 22–5, 36, 56–7, 61–2, 90–2, 102–7, 114–17, 122–4, 130–1, 193; benefits of 114, 116; bilateral 101, 123, 127; continuous 83; developments 106; economic 106, 124; international 77, 128; multilateral 90–2, 96, 100, 102, 106; principles of 75; processes of 6–7, 90–2, 97–102, 106; technical 6, 95, 104–5; in transboundary rivers 114, 130; trilateral 103–4
cooperation dynamics in the Nile Basin 90–108
cooperative arrangements 73–4, 77, 79, 81–3
Cooperative Framework Agreement 6–7, 11, 13–14, 19, 25–31, 33–7, 72–9, 81–2, 91–2, 95, 97, 101–2
cooperative norms, sets of 91–3, 95, 102, 105

Cooperative Water Resources Development 93
costs 46, 49, 54, 79, 81, 83, 114–16, 119–21, 126, 131–2, 158, 160; and benefits 62, 79, 81, 83; dredging 167–8; economic 177; expected total 118; maintenance 132; of non-cooperation 121; political 127; reduction of 117, 121; transaction 8, 108, 130
Council of Ministers of Water Resources of the Nile Basin Countries 25–6
countries 23–5, 28–34, 41–2, 46–9, 51–7, 75–8, 92–8, 100–5, 107–8, 118–22, 126–8, 131–2; dependent 108; developing 147; donor 120; neighbouring 106; stable 97
coupled water systems 140
CPA 97; *see also* Comprehensive Peace Agreement
crops 46, 126, 132, 140, 143, 145–6, 148–50, 160–2, 164, 169; allocation of 150; in Egypt 169; and livestock 126, 132; summer 148; winter 148

dams 5–6, 8–10, 31, 42–7, 54–7, 102–6, 113–14, 118–21, 123–6, 158–61, 167–8, 175–8; disputed 42; of Egypt 9, 79–80, 150; of Ethiopia 10; existing 46, 56; first 31, 42–3; hydro-power 177; joint 116; large hydropower 2, 31, 34, 42, 44–5, 105, 158; multi-purpose 130; projects 2, 45, 52, 102, 116, 128, 130, 159, 161; saddle 45; small 31, 44; of Sudan 43, 119, 126, 168
Decision Support Systems (DSSs) 195
Decision Support Tools (DSTs) 195
Declaration of Principles 2015 5, 74–5, 105, 114, 122, 193
Degefu, G.T. 19, 41, 69
Deltares (Dutch construction firm) 53
Democratic Republic of Congo (DRC) 1, 18–19, 26, 29, 34, 41, 74, 123, 126
Diama project 116
disputes 19–20, 22, 24–5, 31–2, 34, 36, 53, 56, 73, 76, 78, 131–2; bitter 19–20; existing 30; hydropolitical 102; long-standing 28; major 19, 30; peaceful settlement of 22, 24, 32, 53, 71, 73; and treaties 19
distributive justice 7, 61, 63–72, 74, 76–83; application of 74; articulation of 78; central element of 82; determination of 7, 80–1; importance of 65; interpretations of 82; and legitimacy 63, 66, 69, 72; notion of 65, 78–9

Dombrowsky, Ines 5–113
downstream countries 4–6, 10–11, 102–3, 118–19, 125–6, 158–9, 181–2, 184–6, 188–90, 193–4, 200, 210–12; benefits and risks to 160, 194; and concerns over hydrological and environmental impacts 12; demands of 126, 132, 143–4, 199–201, 203; and Egypt's "historic rights" 4, 119, 138; and Ethiopia's "historic rights" 116; and hydrological environmental impacts 12, 83, 119–20, 124, 158, 177, 190; and mean annual releases of water 188; and potential flows 98; and reservoirs 52, 55, 76, 114, 158; and riparians 23, 35, 91, 94, 101, 113; risks to 10, 194, 206; and Sudan's "historic rights" 4, 158, 181; and water flows 187, 189; as water users 51, 200
dredging costs 167–8
drought management policies 139, 200–1, 203–4, 208, 210–11; and GERD 206; to offset the impacts of the GERD 210
'Dutch disease' 161, 176–7

EAPP 123; *see also* East African Power Pool
Earle, A. 98–9
East African Power Pool 123
Eastern Nile Basin 5, 8–10, 62, 77, 93–4, 103, 106, 159, 171, 174–7, 193, 197; hydropolitically complex 90; integration of 127; and poverty alleviation 177; riparians 4; upstream and downstream countries 8
Eastern Nile Council of Ministers (ENCOM) 93, 130
Eastern Nile Subsidiary Action Program 93–4
Eastern Nile Technical Regional Office 72, 90–1, 93, 96, 99–100, 105, 107, 117, 120, 168–9, 195, 198
economic benefits 13, 115, 154, 160–1, 194; to Ethiopia of Blue Nile development 194; generating basin-wide 12, 80, 159, 177; higher Nile-wide 9, 156; potential of 139; shared 7, 13; significant 117; substantial basin-wide 9
economic effects 159–60; of dams using CGE models 159; of the GERD on the Eastern Nile Basin countries 159; of infrastructure development in the Blue Nile River in Ethiopia 160
economic growth 115, 171, 173, 176–7; rates 171, 173, 176

economic impacts 8–9, 12, 154, 158–61, 164, 177–8; of dams 160; to Egypt 154
economists, findings of 79–80
Eddin, Nour 121, 129
effects 29–31, 104–5, 107, 139–40, 167, 169–70, 172–3, 186–7, 197–8, 201, 203, 206–7; on Egypt's economy 140, 152; endowment 175–6; of GERD filling policies on energy generation 139; on irrigation water supply 167, 169–70; on water use in Egypt 169
Egypt 1–6, 8–13, 19–36, 41–57, 73–6, 79–80, 92–8, 103–8, 119–32, 138–41, 167–77, 210–12; and agriculture 194; economy of 8–9, 12, 79, 138–41, 155–6, 161, 173, 177; faces losses in real GDP of US million to US million 171; food industries 36; government of 20, 47, 51, 68, 119–22, 126–9; and hydropower 9, 79–80, 143, 150, 206; and the impact of alternative GERD filling policies 138–56; and irrigation water deliveries 150–1; and the operation of dams 9, 79–80, 150; veto power over any project on the Nile 20; vulnerability associated with a water resource that originates outside its borders 121; water availability 10; water and economic security 140; water supplies 210; water system model 8, 138, 141, 143; water users 203–4, 211
Egypt and Sudan 81, 126; claiming historic and established rights to the Nile 21; demand that provisions on notification of other riparians be included in the CFA 29; indecisiveness regarding attending extraordinary meetings 34; priority arrangement to purchase power generated from the GERD 126; reluctance to endorse the CFA 78; role of journalists 81
Egypt–Sudan Agreement 20, 35
electricity 2, 9, 31, 45–6, 126, 149–50, 154–5, 158, 161, 187, 208
elevations 143–5, 147, 200–1, 204–5, 209–11; critical 206; maximum 201, 211; minimum power production 200, 202, 205–6, 210–11; reservoir 143–5, 147; and seasonal reduction of pool 198
energy generation 139, 166–7, 206, 210; annual 206–7; medium-term 208
ENSAP 93–4; *see also* Eastern Nile Subsidiary Action Program
Entebbe Agreement 25–6

ENTRO 72, 90–1, 93, 96, 99–100, 105, 107, 117, 120, 168–9, 195, 198; *see also* Eastern Nile Technical Regional Office

Environmental and Social Impact Assessment 118

EQL 19, 30, 33, 163; *see also* Equatorial Lakes

equality, principle of 21–2, 35–6, 51, 53, 57, 83

Equatorial Lakes 19, 30, 33, 163

equitable utilisation, principle of 25, 64–7, 71–2, 77, 101, 113

equity, principle of 14, 61, 65, 67, 70–2, 74–5, 78, 82–3

Eritrea 1, 19, 25, 29, 41, 72

Erkyihun, Solomon Tassew 5, 9, 52, 181–90

ESIA 118; *see also* Environmental and Social Impact Assessment

Ethiopia 1–3, 5–6, 18–26, 29–36, 44–57, 68–71, 93–107, 116–19, 121–32, 138–40, 168–77, 206–8; and the Blue Nile 99; decides to increase the capacity of the dam 46, 99, 121; decides to move ahead with the GERD as a national project 91; development vision 117, 159, 177; and the economy 9, 156, 161, 173; ethnically diverse and divided 121; hydropower production and supply 168, 176; and the legal political arrangement with Egypt 107; and the need for exports of hydroelectricity 161; releasing water from the GERD 203; respecting Egypt's historical water share 128; and Sudanese economy 173; and Sudan's integration endeavours 123; and the suspending of construction 121; and the transmission line from the GERD linking the Sudan electricity networks to 118; water resources development and the impact on Egypt's water and economic security 140

Ethiopia Egypt Framework Agreement 1993 22, 24

Ethiopian government 91, 97, 99, 102, 113, 117–19, 122, 126, 129–30; and the capacity to project its economic and political power 97; and the decision to set up a joint military force with Sudan 122; and Egypt 129; and the Grand Ethiopian Renaissance Dam (GERD) 2; and the management of the Dam 126

evaporation losses 18, 42, 45, 56, 120, 140, 144, 147, 160, 168, 171, 173

exports 75, 117, 141, 148–9, 161, 176; agricultural 176; of manufactured products 176

fairness 7, 61–2, 83; and the Blue Nile 68–74; concept of 61–2, 66; and the Grand Ethiopian Renaissance Dam 74–7; and the law of international watercourses 62–8; perspective of 61–83; in relation to legal developments 68

farmers 141, 146, 149–50

FDRE 51, 75, 102, 117, 122–3, 130; *see also* Federal Democratic Republic of Ethiopia

Federal Democratic Republic of Ethiopia 51, 75, 102, 117, 122–3, 130; *see also* Ethiopia

Ferrari, E. 159, 161, 177

filling policies 138–41, 143, 145, 147, 149, 155, 181–2, 184–90, 202; absolute 186, 188–90; effects of 139–40; four-year 187; impact of 140; percentage-based 188; potential 195; unilateral 203

flood protection 200

flow 18–21, 46–7, 53–4, 68, 113, 115, 119, 121, 140, 185, 189–90, 196–7; and the turbines 206; availability of 210; constant 168; downstream sediment 167; excess 211; flood 203; incoming stream 196; intervening 195; minimum 200; monthly 146, 186; peak 118; reduced sediments 168, 194; regulation 206; volume 24, 185; of water and electric generation to Sudan 8; zero released 189; *see also* annual flow

foreign policies 116, 127

framework 10, 24–5, 61–2, 71–2, 76, 78, 83, 114, 116–17, 130, 194; hydro-economic 194; linked water-economic 156; transboundary water interaction 90; wider water diplomacy 194

Framework Agreement 22–4, 35, 78, 92, 100–1

Franck, Thomas 7, 61–7, 74, 81–2

Garretson, R.D. 20, 41, 69

Generalized Algebraic Modelling System (software package) 143, 147

GERD 2–15, 30–2, 44–57, 79–83, 90–1, 95–108, 116–32, 138–43, 163–9, 173–8, 193–8, 200–12; and the agreed annual release 10–11, 202, 208, 210–12; costs and benefits of the project 81; economic impact assessment of 158–77; and

GERD *continued*
Egypt's agriculture 170–1; and electricity to Sudan 47; Ethiopia's decision on 97, 161; filling policies 138–56; and the filling stage 181, 186–8; and financial costs 116, 118–19, 121; and hydropolitics 113, 115, 117, 119, 121, 123, 125, 127, 129, 131; hydropower generation 154; impacts of 80, 140, 156, 173; impounding and operation of the 170, 175; impacting on the economy of Egypt 164–5, 167, 169–71, 173, 175, 177; linking Ethiopia and Sudan electricity networks 118; managing risks while filling the 193–212; negotiations concerning the 14, 127, 130; opening avenues of cooperation beyond water resources 123; operation 80, 165, 169–71, 173–6, 196; project 5, 8, 15, 97–8, 102; reservoir 10, 46, 138, 167–8, 182, 184–5; risks and opportunities for 1–15; rules for the annual operation of the 52, 76, 114, 125–6, 132; seen as a source of political tension 4; and the storage in Lake Nasser 10; *see also* Grand Ethiopian Renaissance Dam
GERDP 51, 54; *see also* Grand Ethiopian Renaissance Dam Project
Global Trade Analysis Project 159, 162–3
Goor, Q. 159–60, 194
government ministers of water resources 25, 32, 49, 51, 54–5, 103–5, 128
Grand Ethiopian Renaissance Dam 2–15, 30–2, 44–57, 79–83, 90–1, 95–108, 116–32, 138–43, 163–9, 173–8, 193–8, 200–12; *see also* GERD
Grand Ethiopian Renaissance Dam Project 51, 54
Great Britain Egypt Agreement 1929 20
green water 29–30
groundwaters 29, 46, 129
GTAP 159, 162–3; *see also* Global Trade Analysis Project
Guariso, G. 113, 158–9, 177, 194
Gueneau, A. 139, 148

Hensengerth, O. 114, 116, 127, 130
High Aswan Dam 11, 13, 20–1, 29–30, 42, 45–6, 106, 119–20, 141, 160–1, 169, 204–5
historical shares of Nile waters 113, 124–5, 128–9, 131

hydraulic infrastructure 4, 90, 92, 98, 159–60
hydraulic projects 93, 115, 120, 130, 132
hydrological conditions 10, 158–9, 164, 167, 195, 197, 203, 208, 210
hydropolitics 32, 81, 113, 115, 117, 119, 121, 123, 125, 127, 129, 131
hydropower generation 3–5, 31, 34, 43–6, 138–41, 143–4, 146, 149–52, 154–5, 160–1, 167–8, 184–7; and Egypt irrigation water deliveries 150–1; in Ethiopia 168; expected 182; and irrigation deliveries 154; and irrigation water supply in Egypt 164; optimal 160; production 130, 143, 149, 155, 159, 168, 177, 198; projects 5, 51, 116–17, 158, 194; and risks to water supply 212; in Sudan 168; timing of 141; upstream 181

ICJ 27, 32, 53, 64–7, 72, 81; *see also* International Court of Justice
ILC 64–5, 67, 75–6; *see also* International Law Commission
inflows 9, 141–4, 147, 164, 181, 196; historical Nile 140; projections 182
infrastructure: developments 107, 159–60; negotiating water and hydraulic 98; new water storage 160
International Court of Justice 27, 32, 53, 64–7, 72, 81
international law 6–7, 21, 23, 28, 32, 51, 61–3, 65, 69, 71, 75–6, 121; context of 62; customary 65, 70, 76–7; principles of 6
international water law 5–7, 21–2, 51, 57, 105
international watercourses 5, 14, 21, 23, 27, 29, 51–2, 57, 62–71, 73–5, 77, 81–3; context of 67, 70; law of 62, 65–8, 77, 83; legal field of 62; management and protection of 57; study of 65; *see also* watercourses
investments 12, 62, 92, 94, 98, 115, 141, 161, 168; cooperative 101; domestic 149; Egyptian 126, 132; foreign direct 96, 106, 168; joint 108; large-scale 97; projects 90, 93, 98
irrigation 3, 5, 34, 43, 94, 126, 128–30, 132, 139, 141, 143–5, 198; deliveries 138; 150, 152, 154; development 99; diversions 198; efficiencies 169, 171, 176; of land 46, 162–5, 176; large-scale national 98; plans 43, 98; revenues 139, 143, 146

irrigation water supply 79, 120, 164–5, 167, 169–70, 173; allocation 170; deliveries 150–1; in Egypt 120, 164, 167; in Ethiopia 169–70; pattern of changes in 170

Jebel Aulia Dam 43, 197
Jeuland, M. 139, 160, 194
Joint Multipurpose Project 90, 93–4, 99, 105, 130
joint studies 52, 76, 78, 81, 103
Jonker, V. 195, 198

Kahsay, Tewodros Negash 5, 9–10, 12–13, 79–80, 119–20, 158–78
Kenneth M. Strzepek 138
Kenya 1–2, 18–21, 26, 29, 33–4, 41, 44, 73–4, 95, 97, 123, 154
Khartoum 18, 22, 32, 48–9, 51, 54–5, 94, 97, 105, 123, 127, 168
Khartoum Document, The 5–6, 14, 32, 35, 54–7
Khashm el Girba Dam 43
Kiir, Salva (President, South Sudan) 122
King, A. 118, 185, 195, 200
Kuik, Onno 158–77

labour 128, 163; demands of 168; unskilled 168, 173–4, 177
Lake Albert 18
Lake Lanoux Arbitration 64
Lake Nasser 8, 10–11, 138, 141–2, 181, 197, 200, 210
Lake Tana 31, 44–5, 197
Lake Victoria 18, 25, 36, 41, 44, 140
lakes 18, 20, 41, 44, 66, 68–9, 141, 184
land 2, 43, 69, 98, 106, 143, 145–6, 148, 150, 159–60, 162; irrigable 162, 164; rain-fed 162–4; usage 145
law of international watercourses 62, 65–8, 77, 83
levels 7–8, 11, 25, 75, 79, 104, 107–8, 145, 148, 150, 169, 173; groundwater 119; minimum operation 186, 200, 203; national 91, 96, 98; reduced reservoir 149, 167

management 4, 57, 81, 91, 93, 98, 100, 113, 123–4, 139–40, 156, 197; cooperative 208; of reservoirs 139
Manantali project 116
Matthews, N. 91, 102, 126
McIntyre, O. 63, 65
Mediterranean Sea 1, 18–19, 41, 140

meetings 25–6, 32–4, 47–51, 53–5, 74, 78–9, 123, 130, 200; annual 33; extraordinary 33–4; high-level 31, 93; trilateral 104–5
Merowe dam 43, 45–6, 56, 97, 167, 169, 201
Mike Hydro Water System Model 142–4, 195
ministers of water resources 25, 32, 49, 51, 54–5, 103–5, 128
Mobasher, A.M. 144–7
model 8–9, 138–9, 143, 146–8, 150, 152, 159–60, 162–3, 177, 184–5, 195–8, 201; partial equilibrium 159–61; water planning 139; water resource 195–6; water systems 8, 138–9, 148
modelling framework 139, 141, 162, 196; dynamic water system-economy 139; global 162; hydro-policy 10, 195–6; hydroeconomic 139; robust analytical policy-oriented 196; to simulate complex multi-objective reservoir operations 10; theoretical 159; well-designed hydro-policy 194
Moges, S.A. 120, 195, 198, 200
Mohamed, K. 119, 129
Mulat, A.G. 120, 195, 198, 200

Nairobi Statement 33–4
National Water Resources Plan 129
NBI 1–2, 11–14, 25–6, 72–4, 90–3, 95–6, 98–102, 107–8, 117, 119–20, 130, 195; activities and projects 96, 99; cooperation processes 13, 100; *see also* Nile Basin Initiative
negative impacts 5, 47, 79, 103, 119; greatest 167; potential 9, 119, 124, 131–2; significant 9
negotiations 5–6, 21, 64, 70, 72–4, 76, 81, 95, 114, 120–2, 124–9, 131–2; benefit-sharing 115; and cooperation 14; guiding state 64; starting point for 124
NELSAP 93; *see also* Nile Equatorial Lakes Subsidiary Action Program
new power imbalances 131
Nile Agreement 1959 27–9, 41
Nile Basin 1–2, 4–7, 11–15, 18–19, 23–30, 34–6, 70, 72–4, 90–3, 104–7, 122–4, 138–40; cooperation dynamics in 90–108; countries (Nile-COM) 26, 33–4, 72; exchanging information on planned measures 29; initiatives and legal frameworks 7; issues 15; key rules and principles 70

Nile Basin Commission 30, 34, 36, 73, 91
Nile Basin Cooperative Framework
   Agreement 18–36
Nile Basin Discourse 81
Nile Basin Initiative *see* NBI
Nile Basin Initiative Act 2002 26
Nile Basin states, and a regional approach 4
Nile Cooperative Framework Agreement 72
Nile Equatorial Lakes Subsidiary Action
   Program 93
Nile River 1–2, 6–8, 18–25, 28–36, 41–5,
   51–3, 56–7, 62, 69–72, 74–8, 81, 140;
   alliance of Egypt and Sudan 48;
   countries 13, 90, 102; disputes 22; and
   fairness in allocation 70; flowing
   through Ethiopia 36; for generation of
   hydropower and irrigation 43; historical
   shares of 113, 124–5, 128–9, 131; and
   riparian states 19–20, 24–6, 29–31, 33,
   35–6, 41–2, 69, 72, 90–2, 95–6, 98,
   129–30; system 27, 29–30, 73, 139;
   system of waters 29; water resources 2,
   57, 69, 92, 100, 107, 113, 121, 123, 139,
   163
Nile River Basin Commission 29, 95
Nile Treaty 6–7, 68–71, 77, 198
Nile Waters Agreement 1929 20
Nile Waters Agreement 1959 20–1, 24,
   32, 35, 41–3, 47–8, 57
Nile Waters Agreement 1991 22
Nile Waters Agreement 1993 22
non-cooperation issues 8, 114–15, 121, 127
northern Sudan 19–20, 56
Nyerere, Julius 20
Nyerere Doctrine 20, 41

Ogink, H. 195, 197–8, 200
Omo River 44
'Opportunities for Cooperative Water
   Resources Development on the Eastern
   Nile: Risks and Rewards' (scoping
   study) 93–4
Owen Falls dam 44

policies 129, 185–8, 190, 195, 200–1,
   203–6, 210–11; eight-year 187–9;
   environmental 162; four-year 187–9;
   fractional 185; operations 105; six-year
   187–8, 190
political changes 91, 96–8
political parties 8, 12, 14, 19–24, 29–32,
   48–9, 51–4, 69–71, 81–3, 115, 193, 196;
   opposition 47, 51; third 49, 116
political tensions 4, 121–3

pollution 23, 63, 67
pool elevations 198, 200–1, 203, 207–8,
   210–11
Port Sudan 127
post-colonial agreements 22
poverty 3, 36, 96, 174
power generation 8, 32, 51, 76, 119–20,
   158, 167, 169–71, 173, 193, 199–200
power plants 45, 171, 173
power production 118–19, 131, 167
principle of fairness, and the law of
   international water-courses 62–3
principles of international water law 32,
   51, 57, 71, 75, 105, 121
processes 6–7, 12–14, 25–6, 61, 63, 67,
   72–4, 78, 81–3, 91–3, 102–3, 105;
   basin-level 13–14; cooperative 132;
   filling 11, 198, 204; multilateral
   cooperative 99; proper 63; trilateral
   102–4
projections 183, 190; of aggregated
   low-frequency value to form 183;
   hydroclimate variability 181–2;
   monthly temperature time series 183;
   plausible 181, 190
projects 4–6, 20–1, 23–4, 44, 93–4, 96–7,
   99, 106–8, 113–21, 123–4, 126–9,
   131–2; benefit-sharing 130–1; capital
   168; controversial 79; cooperative 24;
   dam 2, 45, 52, 102, 116, 128, 130,
   159, 161; designs of 43, 124;
   fast-tracking of 107; hydraulic 93, 115,
   120, 130, 132; hydropower 5, 51,
   116–17, 158, 194; infrastructure 4, 93,
   97, 118, 158; investment 90, 93, 98;
   joint 26, 115, 117, 123; joint
   developmental 127; multipurpose
   investment 106; national 91, 99–100;
   planned 108; potential 14; regional
   interconnection 208
protection 14, 24, 26, 53, 57, 63, 67, 73,
   75, 208, 211; complete 206;
   environmental 29, 61, 90, 93; flood 200;
   of international watercourses 57

Rajagopalan, B. 34, 139, 182
regional cooperation 22, 94
regional institutions 107, 130
regional integration 8, 10, 12, 32, 51, 76,
   82, 115, 123–4, 126, 132
Republic of South Sudan *see* South Sudan
reservoir elevations 143–7; annual average
   145; modelled 146; monthly 144;
   observed 147

reservoirs 9–10, 45, 56, 119, 143–7, 167–8, 181, 184–8, 190, 198, 200–2, 210–12; constructing 140; elevations 143–5, 147; filling scenarios 200; impounding 159, 161; major 196; and meeting downstream flow requirements 198; single purpose hydropower 198

Rieu-Clarke, Alistair 1–15, 61–83

"right process" concept of 7, 62–3, 67–9, 74, 77

riparian countries 4–5, 21–2, 29–30, 42, 90, 92, 100–1, 114–16, 120–1, 127, 130, 190; major 10; sharing benefits 115; upper 101

risks 1, 4, 10–11, 13, 131, 139, 177, 194–5, 203–4, 206, 208, 210–12; of climate change 194; long-term 206; and opportunities 1, 4, 11

River Nile *see* Nile

Robinson, Sherman 5, 8–9, 138–56, 161

Roseires dam 43, 167, 169, 197, 199, 201, 210

rules 11, 23, 52, 62–3, 65, 71, 74, 76, 82, 103, 114, 198; no-harm 27–8; and principles 23, 71; procedural 82; substantive 65

Rwanda 1, 18–19, 25–6, 29, 41, 73–4, 95, 123, 163

Sadoff, C. 4, 114–15, 194

Salman, Salman M.A. 3–6, 11–14, 18, 18–36, 41–58, 66, 72, 74–5, 78, 83, 95, 104–5

SAM 147–8; *see also* Social Accounting Matrix

Shared Vision Program 92–3

sharing: benefits 114, 123, 126, 131; mechanisms 12; procedures 194; water quantities 114; water resources 114, 126

sharing information 121

simulation results 169, 171, 173–4, 176

Social Accounting Matrix 147–8

socio-economic benefits 93–4, 98

socio-economic impacts 103, 118

soil moisture modelling scheme 184

solutions 12–13, 22, 149, 194; cooperative 11; coordinated management 208; equitable 70; fair 80; innovative 194; negotiated positive-sum 14; optimal filling 194; potential 194

South Sudan 18, 30, 36, 41, 95, 97–8, 122

South Sudan Peoples' Liberation Movement 97

stakeholders 10, 82, 193–4, 196

states 4, 7, 21, 23, 26–8, 42, 52–3, 63–78, 81–3, 119–20, 123, 125; affected 27, 52, 75; downstream riparian 23; independent 30, 69, 97; watercourse 63–4, 67, 70

stochastic dual dynamic programming approach (SDDP) 160, 194

storage 10, 23, 34, 45–6, 144, 185–6, 199–201, 203, 210; capacity 2, 31, 42, 44–5, 99, 116–17, 119, 121, 124, 131; initial 185, 210; of rainwater 129

strategies 10, 126, 129, 211–12; cooperative 194; for water resources 129

streamflow 10, 181–3, 186, 189; downstream 188; historical 181; monthly 184–5; reductions in 10; variability 10, 185; volume 185

Strzepek, Kenneth M. 5, 8–10, 12–14, 79–81, 138–56, 159–61, 177, 184, 194, 208

Sudan 1–5, 18–25, 27–9, 31–6, 41–3, 45–53, 55–7, 92–101, 103–7, 118–20, 122–32, 163–77; agriculture 194; and the Blue Nile 103; borders 2, 31, 45, 47, 74; claims 26, 28; complaints 127; consumption expenditure 174; dams 43, 119, 126, 168; diversion structures of the Gezira-Managil 197; economy 173; Egyptian alliance 35; and Egyptian dams 80; and Egyptian power generation 206, 208; electricity networks 118; government 22, 98, 129; hydropower generation 206; journalists 81; ministers 33; Nubians and their relocation 43; President Omar al-Bashir 119; rapprochement 122–3; reservoirs 13, 195, 203, 206, 208, 210; and Uganda 4; water withdrawal options 164

Sudd 18, 25, 34, 41

SVP 92–3; *see also* Shared Vision Program

Swain, A. 21, 116, 118

swamps 18, 41, 184

Tanganyika 20

Tanzania 1, 13, 18–21, 25–6, 29, 34, 41, 73–4, 95, 123, 163

Tawfik, Rawia 53, 90, 97, 113, 113–32, 211

taxes, direct 141, 149

tensions 66, 75, 113–15, 120–1, 126–7, 131; new 130; political 4, 121–3

Tignino, M. 69

Toshka canal 144–5, 200

transaction costs 8, 108, 130

transboundary 55, 103, 159, 161, 178;
agreements 108; cooperation 51, 76, 82,
92, 96, 100, 106, 108; impacts 67, 120,
124, 131; river basins 78, 100; rivers
114, 116, 120, 123–4, 128, 130; water
regime 8, 108; water resources 32
treaties 19–20, 24–5, 28–9, 32, 57, 66,
68–71, 77, 81, 198; bilateral water 28,
66; binding 71, 77; boundary 69;
colonial era 19, 77; design of 66; and
disputes 19; existing 107; new 20; Nile
6–7, 68–70, 77; old 11
tributaries 19, 31, 41–4, 140, 194
tripartite meetings 49–51, 53–4
tripartite negotiations 108
tripartite projects 56
turbines 45, 51, 144, 146, 149, 185–6, 200,
202, 208, 211

Uganda 1–2, 4, 18–21, 25–6, 29, 34, 41,
44, 74, 95, 122–3, 126
UN Watercourses Convention 14, 27, 29,
36, 52, 57, 63–5, 67, 70–1, 73–5, 79
United Kingdom 19, 68; *see also* Britain
United Nations 25, 71–2, 76, 122
United Nations Economic Commission for
Europe (UNECE) 14
United Nations Watercourses Convention
27, 51
United States 197
unskilled labour 168, 173–4, 177
upper riparians 23, 28, 30, 33–6; *see also*
riparians
upstream 10, 57, 73, 93, 100, 108, 185,
190, 210; countries 11, 33, 95–7, 100–1,
106, 117, 126, 131–2, 186;
developments 107–8; hydropower dams
106, 113, 181; infrastructure in Ethiopia
130, 208; riparian countries 4, 21, 23,
27, 36, 91–2, 95, 100–1, 107; water
infrastructure development 102

van der Zaag, Pieter 5, 9, 158–77
'Victoria Nile' 18, 44

water 18–21, 23, 29–30, 46, 66–8, 73–5,
128–9, 138–41, 145–6, 149, 160–4,
185–6; agricultural 150; allocation 21,
29, 114, 124, 126, 169; availability 80,
94, 140; catchments 120; consumption
208; and economic systems 140; and

electric generation 8; and electric
generation to Sudan 8; endowments
169; green 29–30; infrastructure 161;
management 4, 8, 139, 181, 194,
196–7; non-consumptive 197; policies
127, 132; quality 115, 119, 129; rights
92, 114; shares 80, 124–5; sharing 66,
114, 123, 126, 131–2; stagnant 49;
storing 187; stress 148; surface 29;
territorial 69
water flow 8, 23, 119–21, 168; available 8;
downstream 168
water resources 25, 49–51, 54–5, 97, 99,
103–5, 115, 117, 121, 123, 128–30, 198;
common Nile Basin 25, 28, 72; and
foreign affairs 104; management 2, 14,
93, 104, 163; shared 75, 91, 93, 124,
127–8; sharing 114, 126; strategy for
129; systems 160; transboundary 32
water security 6, 27–8, 33, 36, 57, 72–4,
95, 121, 130, 210, 212; defined 27;
downstream 212; national 105;
sustaining 28
water supply 34, 120, 138, 143, 149,
158–9, 167, 175–6, 202, 206, 212;
agricultural 79; continuous irrigation
168; drinking 43; increased 171, 173;
and power generation in Egypt 120;
reduced irrigation 173; reliability
139–40
water system models 139, 142
water systems, coupled 140
water usage 23, 42, 114, 132, 140, 150,
159, 166, 169, 198, 203
water withdrawal 164–5, 169–71; current
level of 169, 171
water withdrawal rates 164, 167, 169, 171
Waterbury, J. 24, 42, 99–100, 117
watercourses 5, 14, 21, 23, 27, 29, 51–2,
57, 62–71, 73–5, 77, 81–3; particular
67, 74; shared 21, 23, 83
Wheeler, Kevin G. 193–212
White Nile 1, 4, 18, 34, 41, 43–4, 140,
195; and Blue Nile 140; and Tekezze 1
Whittington, D. 56, 90, 94, 99, 113, 117,
139, 158–60, 164, 168, 177, 194
Wilson, Scott 193, 195

Yihdego, Zeray 1–15, 20, 61, 61–83

Zhang, Ying 5, 9–10, 12–13, 52, 103, 140,
142, 181–90, 195, 197, 200